Isaac Beeckman on Matter and Motion

NORTH SEA

Groningen

Franeker

Haarlem
Amsterdam

Leiden

The Hague
Utrecht

Delft

Rotterdam

Dordrecht
Gorinchem

's-Hertogenbosch

Zierikzee
Breda

Voere
Arnemuiden
Middelburg
Goes
Flushing

Turnhout

Antwerp

Ghent

Brussels

0   50 km

The Dutch Republic around 1620

Isaac Beeckman on Matter and Motion

Mechanical Philosophy in the Making

KLAAS VAN BERKEL

The Johns Hopkins University Press
Baltimore

The English translation of this book was made possible by a grant from
the Netherlands Organization for Scientific Research (NWO).

The Johns Hopkins University Press
2715 North Charles Street
Baltimore, Maryland 21218-4363
www.press.jhu.edu

Translated by Maarten Ultee. Translation and editorial assistance
from Linda Schneider.

Library of Congress Cataloging-in-Publication Data

Berkel, Klaas van.
[Isaac Beeckman (1588-1637) en de mechanisering van het wereldbeeld. English]
Isaac Beeckman on matter and motion : mechanical philosophy
in the making / Klaas van Berkel.
    pages cm
Revision of the author's thesis under title: Isaac Beeckman (1588-1637) en de
mechanisering van het wereldbeeld.
Includes bibliographical references and index.
ISBN-13: 978-1-4214-0936-8 (pbk.)—ISBN-10: 1-4214-0936-4 (pbk.)—ISBN-13:
978-1-4214-0961-0 (electronic)—ISBN-10: 1-4214-0961-5 (electronic)
    1. Science, Renaissance.   2. Beeckman, Isaac, 1588–1637.
    3. Scientists—Netherlands—Biography.   4. Philosophers—
Netherlands—Biography.   I. Title.
Q125.2.B4713 2013
509.2—dc23
[B]
2012036883

A catalog record for this book is available from the British Library.

Special discounts are available for bulk purchases of this book. For more
information, please contact Special Sales at 410-516-6936 or
specialsales@press.jhu.edu.

The Johns Hopkins University Press uses environmentally friendly book
materials, including recycled text paper that is composed of at least
30 percent post-consumer waste, whenever possible.

Contents

Preface

This book has been long in the making. It goes back to my student days at the University of Groningen in the 1970s, when as a research assistant I was to asked to take a quick look at what Isaac Beeckman's printed *Journal* might reveal about his relationship with René Descartes. The material was so overwhelming that I decided to write a Ph.D. thesis at Utrecht University on Beeckman's contribution to what at that time I called—in the tradition of E. J. Dijksterhuis—the mechanization of the world picture. This book essentially started as a translation, long overdue, of that Dutch doctoral dissertation of 1983. I have, of course, included new material, added new insights, and refined some of the conclusions, but those who are familiar with the original text will see that basically my views of Beeckman have not changed.

Since it has taken so long to produce this study, I have incurred many debts, and it is a pleasure to acknowledge them here. In the first place my thanks are due to both of my supervisors, professor Harry Snelders from Utrecht University and the late professor Edzo Waterbolk from the University of Groningen. The latter was the one who introduced me to the enigmatic figure of Beeckman while I was his assistant at Groningen, and the former invited me in 1978 to do my Ph.D. research at his Utrecht Institute for the History of Science. I will always be grateful for the freedom they gave me in choosing my own topic and for the support they provided me while I was working on the dissertation.

In the years since I defended and published my thesis, many colleagues have encouraged me to publish an English translation. Especially Floris Cohen (Twente University, at present Utrecht University) and Mordechai Feingold (formerly West Virginia Polytechnic, currently California Institute of Technology) never gave up hoping that I would finally succeed in doing so. Yet, if it had not been for Maarten Ultee (University of Alabama at Tuscaloosa), it might never have happened. When I spent some time in Tuscaloosa on a trip through the United States, financed by the United States Information Service, by pure coincidence I met Maarten and out of our discussions grew the plan to try to interest the Johns Hopkins University Press in publishing an English translation of an updated version of my thesis. With generous financial support from the Netherlands Organization for Scientific Research (NWO), which is here gratefully acknowledged, Maarten was able to produce a first draft of a translation of a shortened version of my original publication. In the following years, however, many other things intervened, and the Beeckman book threatened to become one of those legendary books that are always "forthcoming" but never appear in print. So, everything might have gone wrong if Moti Feingold had not invited me to come to Caltech and finalize the book. During two busy months in the fall of 2004, I managed to establish the text of this book. I owe Moti a great debt for not giving up on me and for putting me to work at Caltech.

In the last few years I was able, in between other duties, to put the finishing touches on this book. I thank the director of the Netherlands Institute for Advanced Study at Wassenaar, Wim Blockmans, for giving me the opportunity to do most of this at the NIAS. I also thank Floris Cohen, Christoph Lüthy, and Rienk Vermij for taking the trouble to carefully read through my text and point out mistakes, obscure passages, and misinterpretations. The book has evidently profited from their constructive criticism. In the end, Linda Schneider (Harvard University) thoroughly went through the whole text and not only partially retranslated it but also pointed out many remaining inconsistencies and minor gaps in my argument. I am extremely grateful for the effort she put into it. Nevertheless, only I remain responsible for what I have written about Beeckman and his mechanical philosophy. I hope it will stimulate others to take a look at his fascinating *Journal* for themselves—there is so much more to do.

Isaac Beeckman on Matter and Motion

Introduction

The so-called scientific revolution of the sixteenth and seventeenth centuries is and will forever remain one of the major episodes in modern history. The more we know about it, the harder it becomes to say exactly what happened in the century and a half between Copernicus and Isaac Newton, but we do know that the new science that emerged in that period had a tremendous impact. It changed the way we look at the world, it introduced new ways of interacting with nature, and it even altered the notion of what it means to be human. It set in motion a process of intellectual, social, and cultural transformation that continues to the present day. Understanding the scientific revolution is essential for understanding the modern world.[1]

In the early stages of the scientific revolution, the Dutch natural philosopher Isaac Beeckman (1588–1637) played a crucial but not always recognized role. He was the first to devise a completely mechanical philosophy of nature and thus introduced an approach to nature that was to become a cornerstone of the new science. Beeckman also had a decisive influence on René Descartes, who in the middle of the seventeenth century became one of the figureheads of the scientific revolution. Yet, for all his importance, Beeckman remains a puzzling figure. His life is well documented and his scientific notebooks are available in print (and on the Internet), but there is still no general account of his life and work that allows for a deeper understanding of his contribution to the new science. His originality in devising a mechanical philosophy of nature is generally acknowledged, and his influence on Descartes and others is undisputed, even among French students of Cartesianism, yet the extent to which

Beeckman influenced his contemporaries continues to be a matter of debate. Perhaps simply the fact that Beeckman wrote in Latin as well as in his native Dutch deterred some historians—who otherwise were very well aware of the significance of his ideas—from closely studying his notebooks. As a consequence, our picture of the early stages of the scientific revolution remains incomplete and awaits further scrutiny.

There are essentially two ways to describe and interpret the emergence of the new science in the early modern period. One is to treat the development of modern science as a purely intellectual affair, a process driven by the internal logic of questions and answers. The philosophers' search for the mathematical laws that govern the behavior of natural objects was central to this process. Accordingly, astronomy, physics, and anatomy take center stage in this account of the scientific revolution. The best-known representative of this way of looking at the history of early modern science is the French philosopher Alexandre Koyré, author of the classic *Études galiléennes*.[2]

The second, and opposing, perspective is social. The history of science when viewed in this way is not a set of theories but a bundle of overlapping social practices—ways of doing as opposed to ways of thinking. According to the social perspective, the scientific revolution arose from the practical work of highly skilled artisans, engineers, botanists, alchemists, and medical doctors, and not from discussions among philosophers. In the fifteenth and sixteenth centuries, a new view of manual labor and artificial processes through which nature was altered made its way into the work of artists and the treatises of engineers. Then, during the seventeenth century, the new appraisal of the arts started to penetrate the world of the philosophers. Some of them became convinced that the methods employed by mechanics and artisans to modify and alter nature might also be useful for acquiring knowledge of the natural world. The work of Giordano Bruno, Descartes, Galileo Galilei, and others documents the decisive impact that this new appraisal of techniques and the mechanical arts had on prevailing concepts of nature, philosophy, and science. An early representative of this line of interpretation was the Italian historian of science Paolo Rossi with his *Philosophy, Technology, and the Arts in Early Modern Europe*.[3] He argued that the idea of knowledge as construction, the introduction of the machine model for the explanation and comprehension of the physical universe, and the thesis that man can truly know what he fashions or constructs and *only* what he fashions or constructs were all assertions resulting from the penetration of the philosophical and scientific world by the new artisanal way of thinking.[4]

For many years, the intellectual interpretation of early modern science remained the dominant view. In the 1950s and 1960s, when the history of science as a historical discipline came of age, the opposition between intellectual freedom in the West and social determinism in the communist world heavily influenced the Western image of science—past and present. For example, A. R. Hall, a major historian of science, in 1959 argued that insofar as scholars owed something to the craftsmen, the initiative had always been on the side of the scholars: "It is the philosopher who had modified his attitude," Hall stated, "not the craftsman, and the change is essentially subjective." He pointed out that the success of craft empiricism was nothing new in late medieval and early modern times, and that if scholars began to realize its significance for what they were doing, it was not because such success was more dramatic now than in the past. "It was always there to be seen, by those who had eyes to see it, and the change was in the eye of the beholder."[5]

In recent decades, however, the social interpretation of the origins of the scientific revolution has attracted new followers and by now has become at least as fashionable as the intellectual interpretation. The literature abounds with detailed descriptions of workshops and laboratories where engineers, goldsmiths, potters, botanists, alchemists, apothecaries, and surgeons gathered productive knowledge about the world by dealing with material objects, undisturbed by the scholastic opposition between philosophy and art or between theory and practice. Knowledge was no longer acquired only by reading books but also by the physical encounter with matter. This approach to nature appealed to some sixteenth-century scholars who were dissatisfied with the established philosophy of nature. It dawned on them that the understanding of the material world attained by artisans and others might actually be superior to those of scholars, who only read books. The crafts thus were not only a source of specific techniques and knowledge of materials but also a way of viewing matter and nature. The writings of Paracelsus, Francis Bacon, Descartes, and others articulated, legitimized, and used this new approach to nature to gain new and productive knowledge of the universe and all that constitutes it.[6]

Still, historians of science have only begun to understand how artisanal practices and habits of mind, in short the material understanding of nature, shaped the new science that developed in the seventeenth century. It is one thing to show how the divide between scholars and craftsmen, with their completely separate educational systems and their contrasting social status, was bridged over the years by knowledgeable practitioners and inventive philosophers and how it changed prevailing conceptions in botany, medicine, and chemistry; it is another to demonstrate how the knowledge gained by artisans and craftsmen was diffused in scholarly circles and how hands-on, artisanal knowledge was transformed into the mathematical sciences. A study of a botanical garden or the workshop of a goldsmith may indeed capture the process of creating a body of empirical knowledge, but it still remains to be demonstrated how artisanal knowledge ended up in the highly abstract mathematical constructions of people like Descartes, Christiaan Huygens, and Newton. In-depth studies of natural history, alchemy, and other primarily empirical disciplines are always helpful, but no account of the scientific revolution is adequate as long as we focus on just a few of the many disciplines that constituted the world of science.

For this reason, a detailed study of the life and work of Isaac Beeckman is important for understanding the rise of the new science. Beeckman in a way is the missing link between artisanal knowledge and mathematical science. He is the perfect example of a "liminal" figure between the world of scholarship and the crafts, someone who was actually able to incorporate an artisanal way of dealing with nature into a new and academically acceptable discourse on nature.[7] In contrast to people like Cornelis Drebbel, who was primarily an engineer and craftsman, and Descartes, who above all was an academically trained philosopher, Beeckman was both a craftsman and a scholar. Like his father, he earned his living by making candles and constructing water conduits for breweries and fountains. In the 1630s, he picked up another craft when he learned how to grind lenses. Yet Beeckman was also a theologian, studied mathematics and medicine (though he never practiced), taught logic and rhetoric at several schools, and corresponded with some of the leading mathematicians and natural philosophers in Europe. Artisanal knowledge and scholarly science were united in Beeckman, and he disseminated his way of looking at things in the world of higher learning through his contacts with prominent philosophers like Descartes and Marin Mersenne. His mechanical philosophy of nature was grounded in both the practical knowledge of a craftsman and the theoretical knowledge of a scholar.[8] This is what makes him a crucial figure in the history of the early scientific revolution.

This was *not* a widely held view of those few people who considered Beeckman's role after his untimely death in 1637. Up to the beginning of the twentieth century, he was remembered only as a vague acquaintance of Descartes. Some of Descartes's early biographers did indeed indicate that Descartes, in his youth, had been inspired by this Dutch schoolmaster, but they also made it clear that Descartes had soon outgrown Beeckman's influence and had broken with him when the Dutchman became a little too pretentious. Anyway, no writings of Beeckman had survived and thus his ideas were hard to ascertain. According to Descartes, they were of no particular value.

In 1905, however, Cornelis de Waard, a young physicist and historian of science from the Dutch city of Middelburg, in the Province of Zeeland, discovered Beeckman's original notebooks. In the nineteenth century, these notebooks, long believed to have been lost, had been bought by the Provincial Library of Zeeland simply because the author originated from that part of the country. No one at that time had recognized the notebooks' value, and the library paid the purely symbolic price of half a guilder.[9] De Waard, however, knew of Descartes's remarks on Beeckman and immediately saw the potential impact these notebooks could have on the study of early modern science and philosophy. He set out to transcribe the notebooks and to study specific problems that were discussed in them, like the law of falling bodies and the phenomenon of air pressure. He discovered that Descartes in deriving the law of falling bodies was heavily indebted to Beeckman and that, many years before Evangelista Torricelli and others had done so, Beeckman had demonstrated that a suction pump works not because of some mysterious horror vacui but because of air pressure. De Waard published a few articles about these topics and at the same time tried to get his edition of the notebooks published. Even though the 1930s were difficult times financially, he succeeded in publishing in 1939 the first of a four-volume edition of Beeckman's *Journal*.[10]

Historians of science have long focused on just a few pages in the *Journal*. De Waard and Koyré concentrated on mechanics, Reyer Hooykaas on the relation between science and religion, Floris Cohen and Frédéric de Buzon on music, Mart van Lieburg on William Harvey's theory of blood circulation, and Henk Kubbinga on Beeckman's conception of the molecule.[11] But the wide range of topics upon which Beeckman wrote, studied, and speculated immediately impresses anyone reading the *Journal* from beginning to end. Mechanics, medicine, music, logic, chemistry, and astronomy are only some of the scientific disciplines to which Beeckman devoted his attention. Furthermore, he also used his notebooks to probe his religious feelings, to keep track of the history of his family, and to tell about his interventions in various projects to improve industrial technologies. Finally, the notebooks contain long lists of meteorological observations in several places in Holland and Zeeland and the proceedings of a mechanical society he founded while he was a schoolteacher in Rotterdam. This is not a diary in the ordinary sense of the word, or a scientific logbook, but rather the papers of a highly creative albeit unsystematic natural philosopher, who recounted everything he considered noteworthy.

The *Journal's* apparent disorder constitutes a historian's gold mine. Whereas most natural philosophers can be studied only through their published work and correspondence, in Beeckman's *Journal* one can see how new ideas actually took shape, before they were polished and dressed up for publication. The *Journal* offers unique opportunities to study the philosopher's mind at work. What one encounters here is nothing less than natural philosophy in the making. On the basis of these notebooks, it is possible to finds answers to questions about both the practice and the theoretical context from which Beeckman's ideas arose. These are opportunities we usually do not have in the case of other philosophers.

Yet the value of the *Journal* reaches even further: Beeckman was not just one of the many *novatores* of the late sixteenth and early seventeenth centuries who were dissatisfied with the prevailing Aristotelian philosophy of nature. He stands out because he devised a strictly *mechanical* philosophy of nature, the first of its kind as far as we know. Notwithstanding the disorderly way in which he recorded his thoughts, it is possible to reconstruct the fundamental concepts that underlie his ideas about natural phenomena. Beeckman intermittently summarized the basic principles of his natural philosophy. In his view, nature consists of nothing but matter in motion, and all phenomena can be explained by combinations of material particles that behave according to specific laws of motion. This mechanical way of interpreting natural phenomena lies at the heart of the scientific revolution. Beeckman developed this philosophy without the help of any predecessor and actually appears to be the first natural philosopher to come up with such a philosophy of nature. Now, Beeckman's notebooks—his account of the development of his ideas—allow us to investigate why and how he devised this philosophy. Whereas Descartes had the example of Beeckman—whose ideas he copied extensively—Beeckman had no such example and therefore could only trust his own wits. What then made him interpret the world in the way he did?

With these considerations in mind, it is clear what the agenda should be for gaining a full understanding of Beeckman's contribution to modern science. First, it is necessary to compile a clear and detailed picture of his biographical background. Beeckman did not live a long life, but it was a life full of unexpected career changes that made him familiar with diverse intellectual and social settings. From his early years in Zeeland to his studies of theology and mathematics in Leiden, from his work as a candlemaker to his study of medicine at Caen, and from his involvement in technical projects to his job as a schoolteacher in grammar schools in Utrecht, Rotterdam, and Dordrecht—Beeckman's life was never dull or uneventful. Second, it is essential to reconstruct his mechanical philosophy, which he developed in the 1610s and which revolved around the central notions of matter and motion. Third, there is the question what made him articulate this radically new philosophy of nature. The literature he read, his background as a craftsman, and his religious convictions should all come into consideration, as should his adherence to the ideas of the sixteenth-century French philosopher and pedagogue Peter Ramus. The similarities between the visual strategies developed by Ramus and other humanist pedagogues on the one hand and the tendency of mechanical philosophers to construct visual representations of natural phenomena on the other suggest a link between the classroom and the laboratory, to say the least. The new philosophy of nature of the seventeenth century may very well have been not just a matter of mathematics, mechanics, and experiment but also of mundane pedagogical practices in dusty and stuffy classrooms.

Finally, there is the question of how exactly Beeckman's mechanical philosophy of nature impacted the new science that was emerging in the early seventeenth century. It seems that Beeckman not only influenced contemporary philosophers like Descartes and Pierre Gassendi by suggesting specific solutions for specific problems, such as the derivation of the law of free fall or the mechanism behind the phenomenon of magnetism or the compatibility of atomism with Christian thought. He also presented them with a new kind of natural philosophy in which the mechanic's technical ingenuity is combined with the philosopher's theoretical sophistication. By way of a coda, a further analysis of the picturability that is so central to Beeckman's idea of natural philosophy opens the possibility of showing that this philosophy not only was deeply rooted in the world of craftsmen and schoolmasters but also resonated with developments in the visual arts, thus enabling us to firmly set Beeckman's philosophy in the wider frame of early modern European culture.

CHAPTER ONE

The Making of a Natural Philosopher, 1588–1619

To the modern visitor, Middelburg, the capital of the Dutch province of Zeeland, appears quiet and friendly, perhaps even picturesque. Compared to the bustling city of Rotterdam, an hour away by car or train, Middelburg is a small city frozen in time. The medieval abbey with its cloister and the slender church steeple, the city hall in late Gothic style, and the weather-beaten warehouses along the quiet quays testify to a former golden age, when Middelburg was much more than the provincial backwater it appears to be now.

Four hundred years ago, the atmosphere was indeed quite different. Middelburg, located in the Scheldt estuary close to the thriving metropolis of Antwerp, was then one of the busiest cities in the Netherlands. The northern provinces had just thrown off the yoke of the Spanish king and formed a new state, the Republic of the Seven United Provinces. Holland was the wealthiest and most populous of these provinces, but the province of Zeeland, staunchly Dutch Reformed, shared the general Dutch enthusiasm for maritime power, mercantile ventures, fierce local patriotism, and a fervent desire for self-government at every level of society. Middelburg occupied a central position in Dutch trade during the early seventeenth century. Strategically located, the city served as a crossroads for economic and cultural exchanges between the southern and northern Netherlands.[1]

Middelburg owed its prosperity around 1600 in large part to the influx of immigrants who fled areas under Spanish military control. These southern Netherlanders came north for economic and religious reasons, and Middelburg welcomed them. The city's once-prosperous economy had been badly damaged by war during the years 1572–74. Furthermore, the citizens of Middelburg had lost part of their jurisdiction over the surrounding countryside to rival towns, including Veere and Vlissingen. They even had to reconcile themselves to the emergence of Arnemuiden—no more than an outport of Middelburg—from their control, after its recent acquisition of city rights. Middelburg's citizens therefore gratefully accepted the opportunities presented by the northward stream of refugees from Flanders and Brabant. After the Spanish recaptured Antwerp in 1585, Middelburg sent a flotilla there to pick up those who wished to leave. Between 1580 and 1595, Middelburg registered about 2,500 new burghers, compared to 1,660 in Amsterdam and about 1,000 in Leiden. At least 75 percent came from the southern Netherlands: Middelburg offered the burghers a safe haven without forcing them to abandon Antwerp, which might regain its leading position after the war. Between 1570 and 1600, the population of Middelburg grew from some 6,000 to perhaps 18,000 people; during this period the city had to be enlarged twice, and the area within its walls tripled. The people of Middelburg built more houses, constructed new docks, and erected more impressive walls. Middelburg boomed. The city acquired an unmistakable "southern" accent because so many new residents had come from the south or had parents who were from the southern Netherlands.[2]

The policy of favoring immigration brought substantial economic benefits. In the final decades of the sixteenth century, Middelburg was at one point the second most important commercial city of the northern Netherlands, trailing only Amsterdam. Middelburg was more significant than Rotterdam and Dordrecht, two other major cities in the province of Holland. The traditional trade in French wine from Rouen regained its former prosperity. Hanseatic merchants who had avoided Middelburg for some years returned to the city. The English Merchant Adventurers Company, which had a monopoly in the wool trade, chose Middelburg as a depot for English wool in 1582. Trade with the Indies opened new vistas as the Dutch East India Company, founded in 1602, set up an office in Middelburg. Taken as a whole, the capital of Zeeland around 1600 was a flourishing commercial city with ample employment in numerous enterprises and trades, an attractive place for those who had had hard lives elsewhere.[3]

In its religious life, Middelburg likewise bore a definite southern stamp. Calvinism originally had its greatest appeal in cities in the southern Netherlands such as Ghent and Antwerp, and it had already shown its strength in Middelburg before the outbreak of the Dutch Revolt. But the influx of so many southern refugees made the city and its surrounding region a bastion of the

Calvinist faith. Reformed Church records of Middelburg indicate that 400 southerners took communion there for the first time in 1584; they were joined by 1,155 others in 1585, and 1,150 more in 1586. The rapid progress of Calvinism required increasing numbers of preachers: around 1585 there were six, but their number doubled to twelve by the middle of the seventeenth century. The southern Netherlanders put their stamp on the city's economy and its religious life.

Hendrick Beeckman, a well-to-do burgher from Turnhout in Brabant, was among those who left the southern Netherlands for reasons of conscience.[4] After some years of traveling around Italy, he operated a business in a large house on the Great Market of Turnhout. He traded in candles, tallow, lead, nails, and other wares. He converted to the Reformed faith in 1566 or shortly before. When the weakness of the central government in Brussels that year gave unexpected freedom to the new religion, he surfaced as a deacon in the new Reformed Church of Turnhout. His position became untenable in 1567, however, when the Duke of Alba forcefully restored Spanish order in the Netherlands. Hendrick Beeckman fled to England with a brother and several other family members. He worked as a candlemaker in London, where he died in 1581. Hendrick's second wife and a son from his first marriage, Abraham Beeckman, born in 1563, continued the business in London.[5]

The sudden rush of immigrants from the Low Countries to London caused conflicts with English artisans and merchants. Therefore, many southern Netherlanders who had sought safety in the English capital migrated again, this time to the northern Netherlands. In the early spring of 1586, Abraham Beeckman left his stepmother behind in London and crossed over to Middelburg with two of his underage half sisters. The twenty-three-year-old Beeckman settled in Middelburg as a candlemaker, after taking a burghership oath. He also laid and repaired water pipes in private houses as well as in the many breweries of the city. On January 10, 1588, he married Susanna van Rhee, nineteen-year-old daughter of the wainwright Pieter Jansz van Rhee, who lived in a house on the Beestenmarkt (Livestock Market). Suzanne also belonged to the Calvinist immigrant community. Her father was born in the German town of Jülich, her mother came from Sottegem in Flanders, and the two of them had met in Sandwich, England. Abraham Beeckman, his wife, and his two half sisters then moved into a house next door to his father-in-law—who also owned this house.[6]

Isaac Beeckman was born on December 10, 1588, in this house on the Livestock Market at 10 o'clock in the evening. The name Isaac was an expression of family identity, because it pointed toward the Old Testament God-fearing patriarchs who had left their homeland in order to build a new life. This biblical theme continued in the birth names of subsequent children. The second son, born in 1590, was named Jacob; and the third son, who followed three years later, was called Daniel, a reference to the Babylonian Exile of the Israelites. Isaac Beeckman grew up in a southern immigrant milieu which identified strongly with the Chosen People of the Old Testament, a comparison frequently made in Reformed circles.

BEECKMAN'S EARLY YEARS AND EDUCATION

As the family expanded, the house on the Livestock Market, which Isaac Beeckman later recalled as a "small little house," became cramped.[7] After the birth of Daniel, the family moved to a new house that father-in-law Van Rhee had built in the nearby Hoogstraat (High Street). In that house, named "The Two Roosters," seven more children were born: Susanna (1595), Janneke (1597), Sara (1600), Maria (1602), Gerson (1604), and the twins Abraham and Esther (1607). Abraham Beeckman's two half sisters, Sara and Elisabeth, moved out when they married: the first in 1595, the second in 1600. Meanwhile, Abraham Beeckman's social standing rose considerably. In 1597, when Van Rhee and his wife died of the plague, he inherited "The Two Roosters" and two years later he bought an adjoining property.[8] He assumed a prominent role within the guild of fat merchants, to which all producers of oil and tallow wares (including candlemakers) belonged. Abraham Beeckman served as dean of the guild four times and as *beleeder* (member of the board) eight times between 1593 and 1621.[9]

Abraham Beeckman was a highly cultured man in his own way. Isaac Beeckman later described him as a small, stocky man with a loud voice and a large head, which in his son's view indicated great powers of imagination and a sharp understanding.[10] An eighteenth-century historian said Abraham was "a man well versed in Theology and many languages."[11] Understandably, he also wanted his sons to be well educated. And Isaac Beeckman made good use of his education. At age eleven, Isaac wrote poems and completed a play featuring four characters.[12] In 1601 Isaac was sent to study with Anthonie Biese or Biesius, rector of the Latin school in nearby Arnemuiden, a surprising move. Middelburg had a Latin school whose rector, Jacob Gruterus, was a renowned scholar. Biesius by contrast was relatively unknown, and his one-man school must have paled by comparison with the great school of Middelburg.[13] It appears, however, that Abraham had a theological conflict with the ministers of Middelburg, who were also directors of the Latin school and therefore influenced school matters. As a consequence, Abraham Beeckman preferred to send his son to school in Arnemuiden.

The conflict arose in 1597 on the question of whether children of parents who remained Roman Catholic could be baptized in the Reformed Church. The official teaching of the Reformed Church allowed this practice, but Abraham Beeckman took a stricter view and thought that Reformed baptism should be refused in such cases.[14] He tried to win over various ministers, who were friends, but they refused to adopt his opinion, arguing for church unity above all; some ministers even tried to convince him that he was mistaken. Philip van Lansbergen, a preacher in Goes and friend of Beeckman since his youth, wrote him letters nearly every month in 1600, trying to persuade him of the error of his ways.[15] But Beeckman did not change his views. According to Isaac, although his father possessed a clear understanding, he was also domineering, obstinate, and stubborn. Abraham Beeckman was not a man to suffer correction by a few preachers; at the same time, the ministers viewed him as a troublemaker who attended church only to catch them in error.[16]

Beeckman and his wife were excommunicated in 1603 as a result of these increasingly strident arguments; they could return to the church only if they admitted they were wrong. Beeckman's estrangement from the Reformed community deepened. According to one source, some hundred church members left the church with him. There were numerous opportunities to attend church elsewhere. While Beeckman's daughter Maria was baptized as usual in Middelburg in 1602, his next child, Gerson, was baptized in West-Souburg in 1604.[17] Most of the time Beeckman attended services at the English church in Middelburg.[18] He sympathized with the separatist religious community of Brownists, who formed a small but very active group in Middelburg. Under the leadership of Robert Browne, they had separated from the Church of England in 1581. The Brownists not only objected to many of the Catholic usages of the Anglicans but also viewed the Puritan movement in the Anglican Church as insufficiently radical. They rejected any authority higher than the community of the faithful, be it pope, bishop, or minister; holding such views, they could not remain in England. In 1582 they emigrated to Middelburg, where they held their own religious services in private houses, and according to reports, some Middelburgers also attended. The Brownists survived as a small, hard-core group even after Browne's death in 1583. Abraham Beeckman's attraction to the Brownists was understandable, given that they held similar theological positions (e.g., on the sensitive point of baptizing children of Catholic parents). Isaac Beeckman later developed a close relationship with the Brownist community.[19]

Abraham Beeckman's break with the Reformed Church of Middelburg was not permanent: in 1607, he reconciled with the church council, and as a sign of goodwill, it removed his name from the excommunication records. Beeckman's son Isaac had largely completed his schooling by 1607. After a year at the Latin school in Arnemuiden, Isaac moved in 1602 to Veere, where Biesius had become rector of a new Latin school. Veere was a small seaport town of some thirty-five hundred inhabitants northeast of Middelburg. Biesius had followed the custom of taking along his boarding students, including Isaac, when he moved.[20]

In 1604 at Veere, Isaac started a collection of quotations and aphorisms: *Loci communes*, as he wrote on the title page. Such a commonplace book was the quintessentially humanist method of reading and storing information.[21] Humanists and pedagogues like Erasmus and Vives advised students to select passages of interest for future use, copy them out in a notebook, and group them under headings to facilitate later retrieval and use in prose of one's own. Thus, Beeckman took a bundle of folio-size paper, divided the pages into two columns, and wrote the headings on top of each page. His categories covered the entire range of contemporary knowledge including, "God," "Belief and Religion," "Navigation," "Agriculture," "Riches," and "Metals." However, at fifteen Beeckman was rather old to start a commonplace book, and perhaps that is why he did not fill his notebook after initiating it. The pages remained blank for the moment, with the exception of a quotation from Caesar's *De Bello Gallico*.[22] Nearly all the entries in this notebook, which evolved into his *Journal*, were made at later dates.

Biesius must not have been such a bad teacher after all. At the time of his death in early 1607, his school at Veere had more pupils than the school at Middelburg.[23] It is quite possible that Abraham Beeckman had considered the master's pedagogical skills when choosing in favor of Biesius's Latin school. When Isaac Beeckman went to the University of Leiden to study theology shortly after Biesius's death, he had a solid grounding in classical languages.

Beeckman did not necessarily choose to study theology at Leiden because he wanted to become a minister. According to the biographical sketch later recorded in the *Journal* by his younger brother Abraham, his father had allowed his sons, who were four in number, to take up studies, yet without insisting that they should make a living thereby. "Also he thought that they would not be encouraged to follow the clerical profession because he was a private person without recommendations and the ministers were opposed to him because of some . . . points touching baptism."[24] On May 21, 1607, Beeckman enrolled at Leiden as a student in arts and philosophy; and on October 1, his brother Jacob also enrolled at Leiden.[25] The study of arts and philosophy served as a preparation for one of the higher faculties, and many students took only one or two propaedeutic courses before embarking on one of the higher studies. Isaac and Jacob Beeckman, however, took these introductory courses very seriously and restricted their theological studies to a minimum.[26]

The Beeckman brothers had very good reasons for doing so. In those years, the theology faculty of Leiden University went through a troubled period. When Isaac and Jacob Beeckman arrived, a conflict already had been underway for some years between the more orthodox theologians led by professor Franciscus Gomarus and the more moderate party led by his colleague Jacobus Arminius.[27] The doctrine of predestination was a particular bone of contention. Whereas Arminius argued that the ideas of God's sovereignty and man's free will could be reconciled and that the individual could in some way actively contribute to his own salvation (or even reject God's saving grace), Gomarus stressed man's utter inability to influence the choices God had made beforehand. He maintained that God from times eternal had already elected those who would be saved. Arminius died in 1609, but another professor of theology, Petrus Bertius, continued the debate and managed to have Arminius succeeded by Conrad Vorstius, a German theologian, accused of harboring heretical ideas by the stricter Calvinists. The atmosphere deteriorated rapidly, especially when the conflict took on a political dimension. In 1610 supporters of Arminius went to the States of Holland with a remonstrance, urging the States to intervene, to assemble a synod, and to promote tolerance in matters of faith. A year later, their opponents prepared a counter-remonstrance (thus the Arminian party became known as the Remonstrants, and the Gomarists the Counter-Remonstrants). These conflicts undermined the study of theology in Leiden. The Beeckman brothers, sons of a dissident father, may not have welcomed the prospect of getting involved in yet another theological controversy.

Personal preference also played a role in Isaac Beeckman's choice of studies. He was very interested in mathematics. In the summer of 1607, he took a three-month course from a distant relative in Rotterdam, Jan van den Broecke, "in order to learn a bit of arithmetic, geometry, and navigation."[28] Van den Broecke was one of the many mathematical practitioners who offered their services to the mercantile class in Holland and who were partially responsible for bridging the gap between scholars and craftsmen, which was so essential for the rise of the new science.[29] In the fall of 1607, after spending some months with Van den Broecke, Beeckman went to see the Leiden professor of mathematics, Rudolph Snellius, to ask for advice about how he could deepen his knowledge of mathematics. Snellius gave him a list of books that he thought suitable for self-study.[30] Because Snellius was an adherent of the ideas of the French mathematician, logician, and pedagogical reformer Petrus or Peter Ramus, the list was heavily based on Ramist principles. In this way, Snellius significantly influenced Beeckman's scholarly development.

Beeckman immediately set to work, and in the winter of 1607–8 he pored over his mathematics books until late at night. He collected remarks about mathematics and related subjects in the commonplace book he had begun in Veere. He studied so intensively that after three months he suffered a mental breakdown caused by overwork. He was temporarily forced to give up his studies.[31] The severity of his breakdown cannot be ascertained from the scanty sources available, but Beeckman himself later remarked that in the course of his studies he became a different person. If he had been rather pugnacious in his youth, in later years he was extraordinarily modest, perhaps even shy, and in debates he was always inclined to value the opinion of others higher than his own.[32] In 1608 he returned to Middelburg for a time and occupied himself with other matters, perhaps helping his father in his workshop. Isaac was in Amsterdam in the summer of 1608 with his brother Jacob, who was strongly interested in Hebrew; both received lessons in that language from Henry Ainsworth, leader of the Brownist community there.[33] At Ainsworth's request, Isaac Beeckman at times even preached to the Brownist congregation. He preached in Latin, because, although he could understand English, he could not speak it.[34]

In 1609 Beeckman resumed his studies, and on September 29 he and his brother again enrolled as students in the arts.[35] They both lodged with a Flemish immigrant, Jacobus Brutsart, on the Noordeinde (North End) in Leiden. Isaac then completed his studies without difficulty, and when he left Leiden in mid-1610, he might be viewed as having graduated. It was not customary for theology students to take a university examination. Admission to the ministry was not determined by the university but by the classis, a regional governing body of the Reformed Church that administered a church examination to the candidate and required only a certificate from the university stating that the candidate had attended lectures.[36] Thus, there was no formal conclusion to

Beeckman's studies in Leiden. This also applied to his brother Jacob. Jacob was mainly interested in the study of ancient languages—Greek, Latin, and Hebrew—but around that time many complained because the theological faculty of Leiden was so concerned with dogmatics and exegesis that it neglected the study of classical and oriental languages.[37] Jacob therefore decided to enroll at the University of Franeker in Friesland in September 1610 in order to deepen his knowledge of Hebrew. He attended lectures at Franeker given by the renowned Hebrew scholar Johannes Drusius.[38] Isaac, however, returned to Middelburg, where, with an attestation from Leiden, he first partook of the Lord's Supper on August 8, 1610.[39]

Abraham Beeckman had never intended for his sons to become preachers, and therefore Isaac and Jacob were free to pursue other careers at the conclusion of their studies. While Jacob continued his studies at Franeker with the goal of securing a post in a Latin school, Isaac learned his father's trade of candlemaker. He also worked for his father constructing water supply systems, which he called his "negotia mechanica," his mechanical operations.[40] In less than a year, Beeckman qualified to set up in the trade on his own. On March 7, 1611, after passing the mastership examination, Isaac was enrolled as a new burgher in the city of Zierikzee, the second largest city of Zeeland with probably some six thousand inhabitants. Beeckman joined the candlemakers guild. Later that month, Beeckman was registered as a member of the Reformed Church. In June 1611, Jacob Beeckman also settled in Zierikzee, where he was appointed vice-principal of the local Latin school on a three-year provisional basis. The two brothers, who were very close, probably lived in the same house.[41]

Isaac Beeckman had not yet abandoned the idea of attaining the position of a preacher somewhere, and a year after he settled in Zierikzee he tried to qualify for the clerical profession. In the spring of 1612, he went to Middelburg and from there traveled to Rouen and onward to the French town of Saumur in Anjou, the site of an internationally known Huguenot Academy.[42] At the academy, Beeckman also combined theological studies with the study of mathematics and philosophy; for example, he studied the telescope, an instrument with which he probably was familiar already, because it was invented in his hometown of Middelburg in 1608.[43] In September he passed through Paris, London, Amsterdam, and Middelburg on his return to Zierikzee. In November 1612, Beeckman began to make serious efforts to become a preacher. In July 1613 he underwent an examination by the classis of Walcheren (the island where Middelburg is located). After passing the exam, he was allowed to preach by the classis of Schouwen, where he was better situated to take advantage of preaching opportunities in view of "the condition of his affairs."[44] Thus, while he worked during the week in his candlemaker shop, he preached on Sundays in one of the nearby villages. His brother Jacob, who had already passed the examination in February 1613, likewise combined school teaching and preaching. But both waited in vain for a call to a parish, apparently because none of the Reformed communities wanted to have one of Abraham Beeckman's sons as their preacher. In the end, Isaac and Jacob accepted their situation: they could still become respected citizens of Zierikzee as a candlemaker and a deputy principal.

Making candles did not satisfy Isaac Beeckman: it was a fairly simple craft, which the candlemaker could easily leave to an assistant.[45] Beeckman's assistant Joos Lambrechtsen, a distant relative who had recently come from Flanders and had learned the craft from Abraham Beeckman in Middelburg, did much of the work. Isaac Beeckman devoted far more time to the construction, improvement, and repair of water systems for breweries and fountains for private gardens, which brought him into contact with brewery owners and regents. Sometimes he made "business trips." He traveled to Leiden and Den Haag; in March 1615 he was constructing fountains in Brussels (it had become much easier for northerners to travel to the southern provinces after the Dutch and the Spanish king concluded a truce in 1609).[46] In Haarlem, he studied the system used by local brewers to pipe good, pure water from the dunes to their breweries over the polluted city canals. For this work he stayed in close contact with his father in Middelburg, and Isaac may well have gone on business trips for him, too.[47]

The making of candles, the construction of water conduits, and the repair of pumps inspired Beeckman's inquisitive spirit. We can see the results in his notebooks (the original *Loci communes*): he recorded everything that occurred to him or struck him as noteworthy well after he had completed his studies. During his years in Zierikzee, he was preoccupied with practical questions such as improvements in the production of candles, reflections on the expansion and contraction of fluids, and ways to increase the efficiency of pumps. His research on the speed of water flowing out of a hole in a pipe almost brought him to the formulation of what is now known as Torricelli's law, though he was not aware of it.[48] Beyond such practical matters, Beeckman also made meteorological observations from the tower of the main church of Zierikzee,[49] noted the price of plums, and wrote reflections on the origin of rivers, the squaring of the circle, the nature of light, and the measurement of distances at sea. Furthermore, he incorporated the report of a resident of Zierikzee who had made a trip to the East Indies between 1606 and 1610.[50] Beeckman was not studying systematically in these years. He regarded his student years as over and studied only when he felt like it. Later, he acknowledged he had wasted many hours in this period that he could have used for studies.[51]

There was little stimulus for systematic study in Beeckman's immediate surroundings. Owing to the silting up of canals, the shift of trade routes, and the absence of immigrants from the southern Netherlands, Zierikzee had declined from a busy commercial center to an inward-looking provincial town. Its intellectual and cultural life had far less brilliance than that of Middelburg.[52] But in these years Zierikzee was an important center for Pietism, an orthodox movement in the Reformed Church that strove for a closer agreement between doctrine and lifestyle. Because the doctrines of the faith had been reformed, the Pietists believed emphasis should be placed on applying the new teachings to everyday life.[53] The preacher Willem Teellinck, whose father was a Zierikzee regent, was one of two leaders of this "Further Reformation" with Zierikzee connections. After earning his doctorate in law, he came under the influence of English Puritanism and studied theology as well. Having served as a minister for several years in Haamstede near Zierikzee, Teellinck left in 1613 for Middelburg, where he became a celebrated preacher and writer of edifying tracts.[54] Another "Further Reformation" leader, Godefridus Udemans, who originally came from Bergen op Zoom in Brabant, was a minister in Zierikzee. Udemans was less mystically inclined than Teellinck and wrote more practical ethical works, pointing the way toward a true Christian life for rough seamen and ill-mannered pilots.[55] Jacob Beeckman got along excellently with Udemans and collaborated with him in 1616 to produce a rhymed version of the biblical Song of Songs.[56] Isaac Beeckman, on the other hand, was more introverted than his brother and felt more at ease with the mystical Teellinck. Isaac regarded him more or less as his spiritual guide and found his spiritual home in Teellinck's preaching.[57]

The two brothers were so close that when Jacob Beeckman was appointed principal of the Latin school in Veere on the island of Walcheren,[58] Isaac transferred his shop to his assistant Joos Lambrechtsen and moved to Middelburg, an hour's walk to Veere, in May 1616.[59] Isaac visited Jacob so often that his younger brother Abraham later thought Isaac had lived in Veere, too. Beeckman returned to work in his father's business, where he occupied himself mainly with the installation and maintenance of water pipes, both in Middelburg and beyond. Beeckman again traveled on business, and in the summer of 1616 he went to England "in order to sell pipes."[60] In April 1618 he first went to Brussels to inspect a fountain he had constructed there in 1615,[61] and somewhat later on he delivered an order to a customer in Antwerp.[62]

BECOMING A NATURAL PHILOSOPHER

Beeckman combined his itinerant existence with renewed studies, especially in medicine. He chose medicine as a way of studying nature systematically. He was more interested in anatomy and physiology than in the clinical disciplines. He undertook his medical studies in a systematic manner, abandoning the casual attitude with which he had studied nature in Zierikzee. He made various rules to guide his self-study program. In reading classical authors, for instance, he resolved not to avoid criticism, and he would also make regular excerpts from the books he read. However, he kept these excerpts clearly separated from his own ideas and insights, which were to be noted in a separate book. And Beeckman did as he said: along with his book of excerpts (now lost), he kept filling what he called his "idea book," by which he meant the *Journal* he had used since his student days in Leiden.[63] However, this book was more than just a natural science diary. When he neatly copied a few draft remarks in the *Journal*, he noted: "As I am copying this, I must remember to put the age of all our sisters and brothers in my chaos [in minen chaos], and besides such things that I wish should always be kept, which may be useful to their children."[64] Isaac, as Abraham Beeckman's oldest child, thus assumed the role of family historian who inquired about the lives of nieces, nephews, and ancestors, near and far. And so, the *Journal* also functioned as a family chronicle. Finally, Beeckman wrote advice on how to live the good life, transforming the notebooks into a sort of guide for his life as well.[65] Truly, Beeckman's *Journal* was "a chaos."

Beeckman chose the works of Jean Fernel, a notable French physician from the first half of the sixteenth century, as the basis for his study of medicine, before embarking on a detailed study of Galen (the Roman physician from the second century A.D., whose many works defined the discipline of medicine until the early modern period).[66] Fernel's chief work, the *Medicina*, published in 1554, had become the standard textbook in medical theory for more than a century. Fernel did not assume any extensive anatomical experience and stressed physiology.[67] The book was regarded as suitable for a person who needed to acquire medical knowledge primarily from books. Beeckman "devoured" Fernel's books and then studied Galen, first in Latin, later in Greek. Beeckman never regretted that he had acquired his medical knowledge largely through self-study and had relatively little free time for his studies. Someone who has a great deal of free time, he wrote, is more likely to become lazy than learned, while someone forced to overcome obstacles becomes more diligent and disciplined.[68] He was proud, perhaps too proud, of his status as an autodidact: he had no teachers except for Jan van den Broecke in Rotterdam, who had given him some lessons in mathematics.[69] Strictly speaking, that may be correct, but in Leiden Isaac had received good advice from Snellius. It is also likely that he received books and advice from one of his father's friends, the minister and astronomer Lansbergen. Lansbergen no longer lived in Goes, on the island of Zuid-Beveland, but had moved to Middelburg in 1613.

In early August 1618, after two years of hard work, Beeckman sailed in the company of several others from Middelburg to Le Havre, and from there to Caen in Normandy, arriving on August 7. The University of Caen was well regarded among Dutch physicians.[70] On August 18, he completed his combined baccalaureate and licentiate examination, and received his M.D. with a thesis entitled *De febre tertiana intermittente* on September 6. Beeckman gave a short oration, before his dissertation defense, sketching the place of medicine in the sciences as a whole and arguing in particular that medicine and mathematics were closely related.[71]

Beeckman's dissertation was mediocre. Only one of the two hundred copies that Beeckman had printed has survived, and the surviving copy lacks four central pages.[72] The dissertation consists of twenty theses on tertian fever (a kind of malaria), five *corollaria* (propositions on varied subjects that Beeckman was prepared to defend), and two *quodlibeta* (theses that were opened up for discussion without the defendant committing himself to one of the alternatives). These quodlibeta deal with the Copernican system and the squaring of the circle. The middle pages are missing, but we can fairly assume that in the theses Beeckman gave a logical analysis of the concept of *tertian fevers*, somewhat in the style of Ramist logic that he had learned from Snellius in Leiden, perhaps accompanied by some speculations about the origin of the illness.[73] The corollaries, which have been preserved, were more significant (see fig.1).[74]

In the first of these corollaries, Beeckman argued that the tendency of water to rise in a closed tube was not caused by the "abhorrence of a vacuum." In his opinion, air pressure was the true cause. In the second proposition, Beeck-

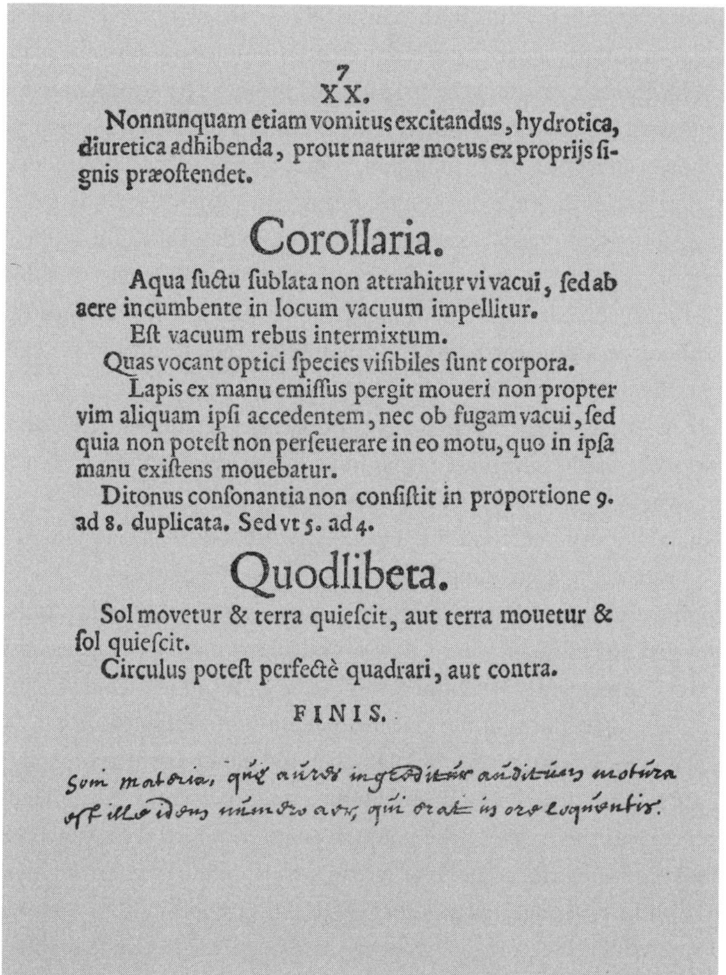

7

XX.

Nonnunquam etiam vomitus excitandus, hydrotica,
diuretica adhibenda, prout naturæ motus ex proprijs fi-
gnis præoftendet.

Corollaria.

Aqua fuctu fublata non attrahitur vi vacui, fed ab
aere incumbente in locum vacuum impellitur.
Eft vacuum rebus intermixtum.
Quas vocant optici fpecies vifibiles funt corpora.
Lapis ex manu emiffus pergit moueri non propter
vim aliquam ipfi accedentem, nec ob fugam vacui, fed
quia non poteft non perfeuerare in eo motu, quo in ipfa
manu exiftens mouebatur.
Ditonus confonantia non confiftit in proportione 9.
ad 8. duplicata. Sed vt 5. ad 4.

Quodlibeta.

Sol movetur & terra quiefcit, aut terra mouetur &
fol quiefcit.
Circulus poteft perfecte quadrari, aut contra.

FINIS.

*Som materias qui auref ingeditur audituus motira
off illo idens numero aor, qui erat us ore loquentir.*

Figure 1. Final page of Beeckman's 1618 thesis (*Journal*, 4:44)

man postulated the existence of a "vacuum intermixtum," pockets of empty
space between adjacent particles of matter. At that time, it was bold to advance
this proposition, although it was not new. His third proposition centered on
the corpuscular nature of light, and the fourth proposition treated the motion
of a thrown object. Beeckman argued against the Aristotelian teachings on
motion that a projectile continues its forward motion after leaving the hand
through the working of the medium, or that it is caused by an added force or
impetus. In Beeckman's view, the object remained in motion simply because it

was not able not to continue in the state it was in when it was moved forward by the hand. Finally, the fifth corollary treated a minor point in the science of musical harmony. Certainly the first four of these propositions were hard to reconcile with the accepted Aristotelian philosophy of nature. The fourth corollary, however cryptic its formulation, comes close to a modern understanding of inertia. Denys Porée de Vandes, Beeckman's supervisor, challenged exactly this fourth corollary.[75] Yet apparently Beeckman skillfully defended his propositions, because the extraordinary praise noted in the register of doctoral defenses suggests that the faculty was very impressed with his talents. In addition to his knowledge of medicine and philosophy, Caen faculty admired Beeckman's mathematical skill and command of Greek.[76]

After receiving his doctoral degree, Beeckman attended the dissertation defense of one of his traveling companions, also at Caen, on September 11, and he made various excursions in the surrounding area.[77] By October 10, 1618, Beeckman had returned to Middelburg.[78] On October 16, he arrived in Breda in West Brabant "to help Uncle Peter work and also for courtship" (Pieteroom te helpen wercken en te vryen oock).[79] "Uncle Peter" was Pieter Cools, who had married one of his mother's sisters and had worked as a tanner in Breda since 1613.[80] Apparently Beeckman went to help out in the slaughtering season, processing the fat that became available from tanning the leather of the newly slaughtered cattle. As a candlemaker and member of the oil and tallow merchants' guild, Beeckman undoubtedly could be helpful. Alas, Beeckman does not reveal the name of the young woman he courted in Breda. He had been looking for love for some time, writing, "from when I went from Zierikzee to live in Middelburg again, I was always troubling myself with lovemaking."[81] It is not possible to determine whether his adventure in Breda involved the woman whom he eventually married.

Breda had been taken from the Spaniards in 1598 by Prince Maurice of Nassau, stadholder of the provinces of Holland, Zeeland, and Utrecht and commander in chief of the army of the young republic. The urban life of Breda, a strong garrison town on the border of the Dutch Republic, was very much influenced by the presence of the army of the Dutch Republic. Many officers and common soldiers from foreign countries served in the army, including a French volunteer, Sieur du Perron. Better known as René Descartes, he was the son of a councillor in the "Parlement" of Brittany.[82] He was born in 1596 in La Haye (present-day Descartes), a small town on the river Creuze halfway between Poitiers and Tours. His family owned a couple of farms around Chatellerault in the Poitou region, and Descartes named himself after one of them, Du Perron. After receiving his education at the Jesuit college of La Flèche and taking a licentiate in law at the University of Poitiers, Descartes was disillusioned with the book learning of scholars. He decided to travel through Europe to acquire real-world experience. In February 1618 he arrived in Breda, where he signed up as an unpaid volunteer with one of the two French regiments in the service of Prince Maurice (France and the Dutch Republic were allied in a common struggle against the Spanish king). There was little for an officer like Descartes to do in a garrison town like Breda during the Twelve Years' Truce, concluded in 1609. He must have spent his time playing cards (at which he was rather good), fencing, drinking, and riding horses, the usual pastimes of soldiers. He was also lucky enough to have a chance meeting with Isaac Beeckman, perhaps the only person in town with whom he could talk about things that really interested him. It proved to be a crucial event in Descartes's life, the pivot on which his whole career turned.[83]

In 1653 Daniel Lipstorp wrote the story of the meeting of Beeckman and Descartes:

It happened in the time that our Descartes was staying in Breda that some mathematician or other . . . had proposed an open competition to all in the city for the solution of a mathematical problem. At the posting place many passersby had assembled, and among them was our Descartes. But because he had arrived in the Netherlands only a short time before, he was not yet a master of the language. Therefore he asked the person standing closest to him (whom he later learned to know as the very renowned Beeckman, principal of the grammar school at Dordrecht and a not inconsiderable philosopher and mathematician) if he could translate the problem for him into French or Latin. Beeckman was willing to respond to this friendly request, and he asked our Descartes to write down the problem and to bring him, Beeckman, the solution. With an eye to this he [Beeckman] gave him his name and address. Descartes did not disappoint him. Because, once he had come home, he solved the problem following methodical rules, with the stone of the Lydians, as it were, with no more difficulty and just as quickly as Vieta had formerly done, when he solved all the difficulties of the problem that Adriaan van Roomen had presented to the mathematicians of the whole world. Afterward, he, Descartes, quickly went to Beeckman to fulfill his promise and show him the solution with the proof. On seeing this, the latter regarded Descartes with respect, admiring his genius as greater than he had expected, treating him in a very friendly manner, and tying lasting bonds of friendship with him.[84]

Many years later, Descartes also wrote to Beeckman that he had not sought contact with him but had encountered him by chance, "in the garrison place where I lived, when I met you as the only one who spoke Latin."[85]

It is difficult to reconcile Lipstorp's account with Beeckman's description of his acquaintance with Descartes in his *Journal*, where there is no mention of an encounter in the street. The first mathematical problem discussed by Beeckman and Descartes, according to the notebooks, has the paradoxical character typical of publicly posted mathematical problems. Descartes tried to prove that in reality an angle does not exist: "Nitebatur . . . probare nullum esse angulum revera." Yet the resolution of the problem certainly required no proof or "constructio," such as Lipstorp reports in his account.[86] Furthermore, Beeckman introduces Descartes gradually into the *Journal*. On November 11, Beeckman first mentions a "Frenchman from Poitou" in the context of a problem that the two of them had discussed the previous day. Descartes is not mentioned again until November 23, when Beeckman identified him as "Renatus Picto" or "Picto" for short, "the man from Poitou," not exactly a sign of great familiarity. Only a few days later, Beeckman calls Descartes by his full name: "Renatus Descartes Picto." Elsewhere, Beeckman referred to him as "Mr. Duperron" or "Mr. de Peron."[87] Finally, a third objection to Lipstorp's tale is its striking similarity to a story told about the sixteenth-century French mathematician François Viète. Lipstorp simply may have invented the story in order to present Descartes as the equal of Viète.[88] Although Beeckman and Descartes probably met by chance, it is doubtful whether this meeting took place in the street while they were reading a publicly posted mathematical problem. They could have met at "Het Vreuchdendal" (the Chamber of Rhetoric), a meeting place for Breda burghers, but also visited by those with introductions, such as Beeckman and officers of the States army. From Beeckman's notes it appears that he attended such literary meetings, and for Descartes such a gathering would have been an attractive change of pace from garrison life.[89]

Once they got to know one another, Beeckman and Descartes found they had many common interests of a scientific nature: the division of angles, the movement of a child's top, the phenomenon of resonance, musical intervals, the hydrostatic paradox, and the centrifugal motion of a stone that is released from a sling (all discussed in the *Journal*). The derivation of the law of falling bodies was the most significant. Beeckman's natural philosophical notions of inertia and acceleration complemented Descartes's mathematical skills. Both men respected one another's talents. Descartes told Beeckman that he had never met anyone who combined physics with mathematics in such an outstanding way.[90]

Descartes honored Beeckman by giving him his first scientific work, the *Compendium musicae* (which incorporated some of Beeckman's ideas) as a New Year's Day present. Descartes had worked on it for much of December, "mea gratia" according to Beeckman (which can mean both "for my sake" and "because I urged him so").[91] On December 31, 1618, Descartes finished his treatise and sent it to Beeckman, who apparently had left Breda. Beeckman received Descartes's present by January 2, 1619, when he was in the port city of Geertruidenberg, on his way to Zeeland.[92] Descartes offered the treatise to Beeckman as "a memento of our friendship and the surest memorial of my affection for you" (familiaritatis nostrae mnemosynon et certissimum mei in te amoris monimentum).[93] He also asked Beeckman to keep the treatise locked in his study and not to show it to others, fearing that others would undoubtedly focus on its imperfections. "Above all," Descartes added, "they would not know that is has been composed for you alone [tui solius gratia], among the ignorance of soldiers by a free and idle man."[94] Beeckman complied with Descartes's request, although many years later he copied the manuscript in his *Journal*.[95]

After Beeckman's departure, Descartes stayed awhile longer in Breda. He finally left Breda in March 1619, traveling via Vlissingen to Middelburg, where he hoped to find Beeckman. Beeckman, however, was traveling in Holland at the time.[96] At the end of the month, Descartes returned to Breda, where he sent Beeckman at least three letters expressing his love and admiration for him. A few days later, Descartes left Breda for good, now heading for Germany via Amsterdam and Copenhagen. Descartes's final letter, as affectionate as the earlier ones, was sent to Beeckman from Amsterdam on April 29, 1619.[97] Beeckman replied with a letter sent to an address in Copenhagen, but we do not know whether Descartes ever received this letter.[98] This brought to an end direct contact between Beeckman and Descartes for the time being. Their contact had lasted from mid-November until the end of April 1619.

Nonetheless these were extremely significant months for Beeckman and Descartes. Descartes credited Beeckman with inspiring him to return to his philosophical studies. On April 23, 1619, he wrote Beeckman:

> If, as I hope, I stop somewhere, I promise you that I shall undertake to put my *Mechanics* or my *Geometry* in order, and I shall honor you as the first mover of

my studies and their first author. For truly, you alone have roused me from my idleness and recalled to me what I had learned and already almost forgotten. When my mind had strayed so far from serious occupations, you led it back to better things. Therefore, if by chance I produce something of merit, you can rightfully claim all of it as yours. As for me, I shall not forget to send it to you, not only so that you can use it, but also so that you can correct it.[99]

Of course these remarks were meant to be flattering to Beeckman, yet is worth noticing that they praise and belittle him at the same time. Descartes claimed the work on mechanics as his own, although, for example, the work on the law of free fall had been their common endeavor; also Descartes portrayed Beeckman as simply an external cause of his new "studies" because Descartes only recalled through Beeckman what he had known all the time. Perhaps Beeckman only saw the praise, but on further review he may have also noticed the first signs of presumption and arrogance for which Descartes is well known.[100]

The meeting with Beeckman was not only important for Descartes's intellectual career. He also was emotionally deeply touched by his new friend. For him, Beeckman, eight years older and much more experienced in life, was not only a mentor, a guide, perhaps a father figure, but also someone to adore. He closed a letter to Beeckman dated January 24, 1619, with the words "Love me and rest assured that I could no more forget you than I could forget the Muses themselves." Descartes felt tied to Beeckman "with bonds of everlasting love." He was not only interested in Beeckman's technical reaction to the *Compendium musicae* but also was anxious to know whether Beeckman progressed in his marriage plans. Descartes expressed concern about Beeckman's well-being and tried to please him. The fact that Descartes wrote to Beeckman with a degree of intimacy that one should normally expect to find between lovers has not escaped biographers. But given the fact that not one of Descartes's letters to others around this time has survived, it is impossible to establish whether such expressions are exceptional and thus indications of homoerotic feelings, as one of Descartes's biographers has suggested.[101] Perhaps it is safer to accept the remarks of another biographer, who relates Descartes's immature expressions of affection to the fact that young men who are educated at residential boys schools (such as La Flèche) have greater difficulty in establishing genuine friendships than men of their age who have not attended such schools.[102]

Beeckman was less enamored of Descartes, although he considered Descartes a friend. He hoped that God would give them considerable time to spend together and "to penetrate to the very core in the domain of studies" (studiorum campus ad umbilicum usque ingressuri).[103] On the other hand, he also behaved like the well-meaning mentor concerned about a younger friend, and so he warned him to be careful. Beeckman ended his last letter to Descartes (who was on his way to Copenhagen and northern Germany) with an intriguing story about a certain Frenchman who boasted to be able to construct fountains from which the water runs perpetually, to know secrets of war and of medicine, and to know ways to multiply the supply of daily bread— although he himself was utterly destitute. "I went to see him, and, when I examined him, he turned out to be completely ignorant of everything, even those things that he was professing publicly. So he won't stand the test [here], and we must send him further north, where the numbskulls welcome illusionists and charlatans."[104]

Was this simply a nice story, or a warning disguised as a joke ("Descartes, be careful, lest you come to be regarded as a charlatan when you do not finish writing the books as you promised"). Or was this a fictionalized version of Beeckman's true impression of Descartes? Although it is hard to imagine that Beeckman meant it this way, Descartes would have viewed the story as an unexpected and painful rebuke by someone whom he regarded as a trusted friend. But, as noted above, we are not sure that Descartes actually received this letter, so further speculation is best left to others.[105]

The difference between Descartes's and Beeckman's reactions to their meeting can easily be explained. For Beeckman, in contrast to Descartes, the encounter was the conclusion and culmination of a certain period in his life, the period in which his conception of natural philosophy had taken shape. The originality of his ideas already was evident at his dissertation defense in Caen in September 1618; the possibilities inherent in those ideas became apparent in his conversations with Descartes three months later. Thus, 1618 was a crucial year in Beeckman's development as a natural philosopher. In the preceding years, the foundations for his natural philosophy had been laid; in the years that followed came the elaboration, refinement, and modification of his views.

CHAPTER TWO

Schoolteacher and Craftsman,
1619–1627

While Beeckman and Descartes discussed natural philosophy in Breda, farther north, in Holland and Utrecht, the great religious conflict of the Twelve Years' Truce entered its final phase. In the summer of 1618, Maurice of Nassau, the military leader of the Dutch Republic, took the side of the Counter-Remonstrants, the hard-liners among the Dutch Reformed. In his official position as stadholder, Maurice was only the servant of five of the seven autonomous provinces that together constituted the Dutch Republic, but now—with the consent of the States General—he seized power. In August he proclaimed a state of emergency throughout the province of Holland and arrested Johan van Oldenbarnevelt, pensionary to the States of Holland and leader of the Remonstrant party. After a show trial, the aged Oldenbarnevelt was condemned to death and beheaded on May 13, 1619. In the meantime, a national synod of the Reformed Church assembled at Dordrecht to settle the religious issues; its sessions opened on November 10, 1618. The synod condemned the Remonstrants' positions and in July 1619 established rules of doctrine to serve as touchstones of the true faith.[1]

Besides their stronghold in Holland, the Remonstrants also had been well represented in the city of Utrecht and thus Maurice's coup d'état brought changes in the Utrecht city administration as well. Some councillors had to retire, all Remonstrant preachers had to leave the city, and teachers at the Latin school were tested for orthodoxy. The principal and vice-principal of the school were fired after refusing to sign the declarations of Dordrecht. In October 1619, Antonius Aemilius, who had previously been principal at Dordrecht, was appointed as the new principal. On November 15, Aemilius proposed appointing "a young man in his thirties from Middelburg, named Ysaacus Beeckmannus" as vice-principal.[2]

A SCHOOL CAREER

Aemilius and Beeckman had studied together in Leiden and strengthened their friendship during a shared stay at Saumur in 1612. Beeckman later referred to Aemilius as "his special friend."[3] Beeckman was appointed vice-principal at Utrecht on November 17, 1619, at the not inconsiderable salary of five hundred guilders (excluding a portion of the tuition money). His main duty was to give lessons to the third class, but his teaching responsibilities also included a special course in cosmology.[4] He received free housing and probably lived in the old Hieronymite Convent, where the Latin school was housed.[5] As he had not yet married, his younger sister Maria accompanied him to Utrecht to keep house. After overcoming his father's objections (his father probably believed he could have found a better position), Beeckman began teaching on December 11, 1619.[6] Henceforth, Beeckman would make his living behind the damp walls of convents turned into Latin schools, after having spent at least eight years as a craftsman.

The Latin school was a form of secondary education. Pupils typically entered the school at age ten or eleven to prepare for the university. Yet the school also served citizens' children who sought only a general education and were not planning to continue their studies. After completing the lower grades, most of these pupils left school; some, for example, left to work in their parents' shops. As a result, classes in the lower grades were considerably larger than those in the upper grades. Supposedly the pupils had already learned to read and write in primary schools (so-called Dutch schools), but in practice the distinction between primary and secondary schooling was not so clear. For pupils who had not completely mastered reading and writing, the larger schools always employed a "writing master." Nor was the line between secondary schools and higher education clearly defined. At some schools, subjects in the higher grades were taught at the university level. In some towns, this later led to the establishment of "illustrious schools," where professors gave academic instruction without having the right to confer degrees. In 1634 Utrecht became such an "illustrious school."[7]

All courses were taught in Latin. In the lower grades, teachers began with pronunciation, vocabulary, and elementary rules of grammar. This instruction became steadily more extensive in higher grades. Subjects such as rhetoric and logic, and sometimes cosmology and history also were taught. But the emphasis still fell heavily on passive and active knowledge of Latin. Theological instruction played only a minor role in the curriculum. The Heidelberg catechism was taught only to children whose parents belonged to the Dutch Reformed Church. Even in the seventeenth century, the Latin school retained more of a humanistic than a Reformed character, although the teachers were carefully examined regarding their religious beliefs.

Classical teaching, not individual teaching, reigned at the Latin school. In each grade, one teacher taught all subjects to all pupils. In 1619 at Utrecht, seven teachers taught eight classes, with the principal teaching the two most advanced classes, known as the *prima* (the highest) and the *secunda*. The Utrecht school enrolled about three hundred pupils on average, which was rather large for its time. Even in winter, the lessons began at seven in the morning. There were two long vacations: one at Eastertime and the other in July and August; students also had a week off between Christmas and New Year's Day. In the vacation periods, pupils from outside the city could return home to visit their families; during the school year, they lodged with the principal.

Beeckman was not planning to stay long at Utrecht. Thus he did not transfer his church membership from Middelburg to Utrecht and did little to make new friends. His only close acquaintances were the principal Aemilius and the writing master Evert Verhaer, who was also organist and precentor of the major church of the city, the Dom. Verhaer, a pupil of the famous Jan Pieterszoon Sweelinck at Amsterdam, gave Isaac several singing lessons, without much success. "My voice," Beeckman wrote later, "is so bad that my master said he had never heard anyone worse."[8] However, he learned to play the harpsichord.[9] He also continued his medical and pharmaceutical studies, adopting a theoretical approach because his eyesight was too poor for practical or empirical research. (Beeckman was nearsighted and later was plagued with cataracts.) He also lacked good contacts among the spice merchants of Utrecht, the spice merchants being the only people who had enough knowledge to enable him to profit by an exchange of ideas.[10]

Beeckman kept in close touch with his family in Zeeland. His father visited his son at least once, when he discussed the flow of water in water conduits of breweries. Abraham Beeckman also wrote letters to Isaac, including one in which he informed his son about the famous submarine constructed by their countryman Cornelis Drebbel in London.[11] Beeckman, in turn, made a lengthy visit to Zeeland between early April and mid-May 1620 for his wedding. On

April 20, 1620, the day after Easter, Beeckman married Cateline de Cerf in the New Church of Middelburg.[12] Cateline was the youngest daughter of a rich Flemish merchant's widow, who had come to Middelburg via Calais in 1615. When Cateline became a member of the Reformed Church in Middelburg in 1618, she lived in a house on New Haven, not far from Isaac Beeckman's family home. He must have become acquainted with her there, after his departure from Zierikzee.

Beeckman was about to return to Utrecht when he became involved in a major hydraulic project that had become the talk of Middelburg. A man called Daniel Noot or Nota, a soap boiler and inventor from the southern Netherlands who lived in Middelburg, had proposed to the city council a completely novel way of deepening the Middelburg harbor.[13] Although an ad hoc committee presented a negative report on the project, Nota did not give up and tried to win support for his plan from several private experts. On May 10, 1620, Beeckman and nearly twenty others willing to evaluate Nota's project attended a presentation. One of the others was the renowned poet Jacob Cats, who had been involved for years in reclamation projects in Zeeuws-Vlaanderen (a stretch of Flanders along the Scheldt that was held by the Dutch) and was known to be an "amateur" of hydraulic works. Beeckman initially favored the project, then changed his mind when he learned more about what Nota proposed. Nota wanted to install a wooden barrier consisting of several containers with locking doors across the entire width of the harbor. At high tide, the barrier and harbor behind it would fill with water. As the tide began to fall, the locks would be closed. The locks would open only when the tide was at its ebb, so that the accumulated water would rush out with great force and flush the harbor clean. According to Beeckman, however, the barrier would not be strong enough to withstand the dashing and pounding of the water inside, just as a grounded water ship could not withstand the force and pressure of water inside the ship. The plan was never carried out, and Nota was never heard of again.[14] This was the first time that Isaac Beeckman was consulted as an expert in a technical project.

Beeckman did not like living in Utrecht and took advantage of an opportunity to leave at the end of 1620. The principal of the Latin school in Rotterdam suffered from dementia, and after a search in November 1620, the town council offered the job to Jacob Beeckman, who was still principal at Veere.[15] Before the matter was settled, Jacob Beeckman informed his brother. The Beeckman brothers had cherished the hope of teaching together somewhere for many years. Their plan entailed Isaac Beeckman assisting Jacob at Rotterdam by teaching the highest classes without pay. They agreed to share Jacob's income—a salary of six hundred guilders, a part of the tuition, and fees from boarding pupils. That income was adequate for the time being. Jacob and Isaac Beeckman were both married but had no children; they only had to support their sister Maria, who lived with them. Isaac Beeckman submitted his resignation to the Utrecht city council on November 9, 1620, gave his last lesson there December 11, and moved "cum tota familia" to Rotterdam on December 20.[16]

Along the Maas (Meuse) in Rotterdam, Isaac quickly felt more at ease. He was more open to all that happened in Rotterdam and took an active part in social life. Rotterdam boasted a rapidly expanding center of trade and industry as well as a more vibrant urban culture than Utrecht. Yet Isaac's happiness in Rotterdam can be largely attributed to his brother Jacob, who enjoyed considerable prestige as principal of the Latin school and involved his older brother in many activities.

The school was established in the former monastery of the Cellites (Cellebroeders), not far from the Great or St. Lawrence Church. The complex housed not only the classrooms but also the residence of the principal and smaller rooms for boarding pupils (about sixty at that time).[17] The school flourished under the joint direction of the Beeckman brothers: it had a good reputation in Holland and drew pupils from far and wide. Some Dutch merchants in England even sent their children to Rotterdam to receive a Dutch education; and some professors at the University of Leiden preferred the school at Rotterdam to schools closer to home.[18] There were numerous pupils from Zeeland, several having come with Jacob Beeckman from Veere to Rotterdam. Isaac Beeckman sometimes discussed (presumably out of class) scientific issues with his pupils. Thus, George Ent from Sandwich, England, who later became one of the first supporters of William Harvey's theory of blood circulation, received part of his scientific education from Beeckman.[19] So did Martinus Hortensius from Delft, who became an astronomer.[20] (As we will see in the next chapter, Beeckman permitted Hortensius to study his notebooks, although not at this time.) Another pupil, Frederik Stevin, son of the engineer Simon Stevin (who died in 1620), enabled Beeckman to see the senior Stevin's unpublished manuscripts. In 1624 Beeckman made several visits to Simon Stevin's widow in Hazerswoude, a village to the east of Leiden. Eventually he even received permission to take the manuscripts home in order to study them at his leisure and copy excerpts.[21]

The rapid growth in the number of pupils soon led Jacob to hire two more teachers in 1621.[22] Finally, when Henricus van Cranenburg, the official vice-principal, gave up his post in 1624, Isaac Beeckman was officially named vice-principal, on November 4, 1624.[23] This paid position (he now earned 450 guilders and free housing) enhanced Beeckman's ability to support his family: his wife had given birth to a daughter, after two sons had died in infancy.

The Beeckman brothers' additions to the traditional curriculum (Jacob taught Hebrew; Isaac, logic and rhetoric) contributed to the Rotterdam Latin school's good reputation. Although logic and rhetoric were part of the curriculum of a traditional Latin school, Isaac Beeckman gave advanced exercises in practical applications to the pupils in the highest class. As late as 1670, the rector and senate of the University of Leiden in their *Short Report on the Decline of Studies (Cort bericht . . . aengaende het verval in de studien)* still cited the education given in "the school at Rotterdam at the time that Jacobus and Isaacus Beeckmanus were principals there" as a model.[24] They furthermore specifically observed that in Rotterdam no one was promoted to the university, "who had not practiced in the reading of good authors and histories, from whom they had to tell in public what was most remarkable, in writing Latin letters, presenting dissertations, giving orations, acting as opponent or defendant. For only through these exercises can one be prepared to come to the Academy and to complete to his entire satisfaction his studies under the guidance and direction of the professors."[25] Thus, Jacob and Isaac Beeckman excellently prepared their students for a university education. Isaac Beeckman also helped some students who had already left the school with their academic studies.

When the States of Holland in 1625 consulted the cities about new education laws that would, among other things, give each Latin school an extra class devoted to the teaching of ethics, physics, arithmetic, and astronomy, it is safe to assume that Jacob and Isaac Beeckman advised the Rotterdam city government to adopt such a requirement.[26] But their own ideas went much further: Beeckman believed it was necessary to break away from the system of class teaching. "In order to make a good school, to raise the children in piety and learning," Beeckman asserted that teachers should take the initiative and assume a stronger advisory role, following the example of English colleges.[27] The independent activity of the pupils would have to be stimulated: older ones should teach younger ones, and the quick should help the slow.[28] Beeckman also considered establishing a Collegium Mathematicum for interested and able pupils, as well as for adults. Beeckman envisioned that under his leadership, "learned" participants in the Collegium would instruct the "unlearned" ones, thus allowing him to serve more people than was possible by normal teaching methods.[29] None of these plans came to pass, but something of the last project can be noted in Beeckman's founding of the Collegium Mechanicum in 1626.

BOTH TEACHER AND CRAFTSMAN

Beeckman's roots lay in the world of artisans and merchants, and he did not distance himself from this background when he entered the world of education. In Rotterdam he was on familiar terms with regents and entrepreneurs as well as craftsmen and small traders. His friends ranged from the prominent cloth manufacturer Gerard van Berckel (a member of the city council from 1618 until his death in 1634, who was chosen burgomaster three times), to the modest silk dyer Jan Weymans, who, like Beeckman, had come from the southern Netherlands. When Weymans's wife died in 1623, Beeckman became one of the guardians of their children.[30] Beeckman intermittently practiced his old trade of laying water conduits. When he was in Zeeland for the wedding of a sister-in-law, he repaired the water supply of a brewery in Veere. The plumbing had been installed only a few years earlier by English contractors, but Beeckman found out that the work had been substandard.[31] Beeckman's interest went well beyond that of a mere tradesman: in his notebooks, he wrote about the relationship between the diameter of a pipe and the velocity of the water flowing in it.[32]

Beeckman developed a reputation in Rotterdam as both a talented teacher and a craftsman. In the early decades of the seventeenth century, Rotterdam underwent an enormous economic expansion. One twentieth-century historian, Cris te Lintum, even spoke of the "American years" of Rotterdam.[33] For much of the sixteenth century Rotterdam had retained the character of a mid-size fishing port. During the Dutch Revolt, however, Amsterdam was temporarily eliminated as a competitor when it remained loyal to the king of Spain until 1578 and suffered a trade blockade by the rebels. Rotterdam was one of several cities that profited from this temporary blockade of Amsterdam. Also, the influx of many rich and entrepreneurial immigrants from the southern provinces strengthened Rotterdam economically by the end of the century. With nearly twenty-thousand inhabitants, Rotterdam had become the fifth largest city in Holland. At the same time, the economic allure of its main competitor in South Holland, Dordrecht, receded significantly. By the time Beeckman settled in Rotterdam, the city had also surpassed Middelburg, whose geographic location appeared increasingly peripheral in the Dutch Republic. Ultimately, the citizens of Rotterdam acknowledged only Amsterdam as their larger and more powerful rival.[34]

The economic expansion led to a dramatic transformation of Rotterdam's external appearance. Its citizens dredged new harbors and built new warehouses, revitalizing the waterfront. As a result, increasing numbers of new industries were drawn to the city. During the Twelve Years' Truce, fifteen beer breweries alone were established, nearly all close to the Maas River, which provided a source of good water. The breweries and their associated malt works, vinegar works, and horse mills represented a substantial investment. Beer brewers were among the richest and most influential citizens. But the textile industry, where immigrants from the southern Netherlands were prominent, also flourished. Gerard van Berckel, the prominent cloth manufacturer, originally came from 's-Hertogenbosch in Brabant.[35]

The industrial expansion of Rotterdam accompanied a modest mechanization of industry, primarily but not exclusively in textile manufacturing. At the instigation of several entrepreneurs from Tilburg in the province of Brabant, for example, the introduction of the fulling mill led to the partial mechanization of the linen industry. Political and economic complications arose, though, as a result: as late as 1637, the city council abandoned plans to construct a municipal fulling mill because of protests from the hand fullers, who were threatened with unemployment. Council members could not even walk on the planned fulling mill site, "being placed in evident danger to their lives."[36] Yet this did not prevent the largest entrepreneurs from continually searching for methods to simplify and mechanize production processes. People feverishly sought new and better technology, and patents were requested for new mills, locks, ships, and new ways of preparing paint throughout the young republic. Few of these inventions resulted in significant advances in manufacturing processes, but the pressure to innovate was striking.[37]

In the summer of 1626, Beeckman learned from Van Berckel that one of his fellow regents, burgomaster Nicolaas Puyck, proposed to pay a hefty sum—forty thousand ducats, Beeckman later reported—for a share in a patent for a greatly improved horse-driven mill.[38] An inventor—no name is mentioned—had submitted an application for the patent and had raised money among Rotterdam merchants and industrialists to finance the construction of the mill. One of the largest beer brewers in the city, Willem Aertsen van Couwenhoven, also had a share in the enterprise. They wanted to use a new type of horse mill in the malt works annex to the breweries, where malt (the germinated barley that serves as raw material for beer) was ground. Expectations were high: supposedly the new horse mill would grind three or four times as much malt as an ordinary mill. The inventor declared that the invention rested on the principle of perpetual motion. This prompted Beeckman to warn Van Berckel that the invention would not work: "Only God makes living gears or perpetual motion."[39] Beeckman implored Van Berckel to warn Puyck as well.

When Puyck heard that the vice-principal had doubts as to the soundness of the invention, he invited Beeckman to come and explain why the mill would not work. On July 15, 1626, Beeckman visited Puyck, inspected the model, and discovered the fault in the machinery, despite seeing it for the first time. Beeckman spoke confidently because he had seen a nearly identical machine two years earlier at his friend Weymans's house. A simple blacksmith had made an apparatus in which two gears turned, claiming that the larger gear by its weight alone would always keep the smaller one in motion. Beeckman had tested it and discovered that it did not work, as he had predicted. Compassion prompted Beeckman to buy the useless invention from the impoverished blacksmith for two guilders.[40] Now, two years later someone else was trying to sell an analogous invention for a lot more money.

Puyck quickly withdrew his financial stake in the enterprise. But the inventor, who was busily setting up a test facility in Van Couwenhoven's brewery, continued his work. On August 6, the apparatus was ready, and the test could be carried out. Curious spectators had come from afar, after Puyck announced Beeckman's predictions that the test would fail. Beeckman himself brought people to witness the experiment, which was indeed a fiasco. At the halfway mark, the brewer removed the malt from the new mill in order to put it in another, traditional mill to complete the process, which worked more quickly and smoothly. The brewer had no more faith in the invention, saying, "Tomorrow I will pound on it with a hammer and take it down."[41] He and several other brewers lost considerable amounts of money. In addition to losing the money they had invested in building the test model, the brewers had also leased less pastureland in hopes of making do with fewer mill horses. Now they had to see whether they could still lease sufficient pasture. This episode, however, definitively established Beeckman's reputation as a technical expert. As a result, Beeckman secured a contract from Puyck to install a fountain in his garden, which he executed with pleasure.[42]

Confronted with so much misunderstanding of the most elementary rules of practical mechanics, Beeckman wondered if it would not be better to coordinate attempts to improve existing technology and evaluate them beforehand with more people, so that disappointments such as with Puyck's horse mill could be avoided. The building of an unsound spire on the St. Lawrence Church in 1621 might also have been prevented.[43] Even the city of Breda might have been saved from capture by the Spaniards in 1625, when its defenders accidentally flooded their own lines twice instead of the Spanish encampment.[44] Thus, in the summer of 1626, Beeckman came up with the idea of organizing a *colle-*

gium or society for experienced traders and craftsmen to meet once a week, or every other week, to discuss problems of practical mechanics. The society's purpose would be to increase technical knowledge among its members. On August 28, 1626, the Collegium Mechanicum—the name Beeckman chose for the society—met for the first time.[45]

THE COLLEGIUM MECHANICUM

The Collegium Mechanicum should be viewed as a study group of like-minded friends. It was not a formal educational institution such as the Collegium Mathematicum, for which Beeckman had also developed plans, nor should it be compared with Gresham College in London, where after its foundation in 1596 free public lectures in mathematics and related subjects were given.[46] Besides Beeckman, there were eight other members of the Collegium Mechanicum, all of whom appear to have been active in trade or industry, or who, like Beeckman, had roots in the world of practical affairs. Nothing is known about "Mr. Pieter," but Dirck Boefie was a corn dealer, Abraham Willemsz a mill builder, Huygh Teunisz a carpenter, Lambrecht Dircksz a shipwright, and Jan Weymans a silk dyer. The only other member with an academic degree besides Beeckman was Johannes Fornerius, M.D., who had begun as a ship's barber and had received his degree only a short time earlier, on July 30, 1626, in Leiden.[47] However, Fornerius did not play any significant part in the Collegium. Medical subjects were beyond the scope of the society, and Fornerius's musical interests (he played the small organ in the St. Lawrence Church and supported the congregation during meditation after the sermon) left no traces in the discussions of the Collegium. Worthy of note, though, is that the two academics in the Collegium both had an artisanal background.

After Beeckman, Jan Jansz. Stampioen Sr., a mathematical practitioner of some repute, was the most important member of the Collegium. Stampioen had begun as a pilot on ships that sailed to Northern Europe, but he taught navigation to ship captains and first mates in Rotterdam since 1617. He also worked as a surveyor, cartographer, and inspector of weights and measures at Rotterdam.[48] Beeckman and Stampioen had known each other only a short time, but they apparently got along well, despite the fact that Stampioen was a Remonstrant and Beeckman a Counter-Remonstrant (Beeckman, although very strict in his religious convictions, always showed a tolerance toward other denominations that was quite exceptional for a man with his background).[49] In June 1626 the two men discussed the question of why shipbuilders always made ship's masts lean slightly backward, and in August Puyck called on them for help in a project to enlarge the surface area of sails for ships.[50] Beeckman and Stampioen's other joint research projects date from the same period, and it is probable that the Collegium Mechanicum grew out of their collaboration.

Members of the Collegium met every Friday at each other's houses beginning on August 28, 1626. Before adjourning, members decided on an agenda for the following week's meeting so each participant could work on solving the problem in advance. The members took turns serving as chairman. Initially, no minutes were taken, and the "common book" (gemeyn boeck) in which their activities were later recorded, unfortunately, has been lost.[51] Beeckman's *Journal* notes are therefore our only source.

A wide range of subjects came under discussion. The members of the Collegium began with the question of whether it was true that sound travels faster horizontally.[52] They also discussed whether an arrow shot horizontally makes a brief upward movement immediately after leaving the bow, before beginning its descent.[53] After these two questions had been answered in the negative, though, they had exhausted their abstract topics. Participants then discussed only practical subjects, such as improvements in the construction of mills, foundations for houses, drying installations in grain warehouses, the deepening of harbors, and chimneys that did not draw air properly.

The subject that received the most attention was the possible benefit of a windmill with sails that turned horizontally, instead of the more usual type with vertical sails. Horizontally turning windmills were not new: in Persia and China they were common, and the Croation polymath and inventor Faustus Verantius had already depicted several models in his first edition of *Machinae novae* (1595). New variations on this old theme were tried amid the general enthusiasm for inventions that captivated the young Dutch Republic at the beginning of the seventeenth century. One of the inventors was a certain Gijsbrecht Pietersz., a master carpenter of Leiden, who received a patent for a horizontal windmill in 1622. His model featured a housing that could be closed and which was designed to protect the moving sails from counter drafts or, in case of very strong winds, to cover the windmill entirely. Three years later, on January 28, 1625, the same inventor, who had moved in the meantime to the village of Rijswijk (between The Hague and Delft), requested a new patent for a horizontal windmill with eight sails. It is quite possible that the same Gijsbrecht Pietersz. had built the models that Beeckman and his friends wanted to study closely. The water-pumping mill with four horizontal sails, discussed by the Collegium Mechanicum and first noted by Beeckman in April 1625, had indeed been set up near Rijswijk. The question was whether such a mill, which Beeckman called a "merry-go-round" (in Dutch, "mallemolen"), "is good and suitable for the profit of this land or not."[54]

Horizontally turning windmills were supposed to have many advantages. A vertically turning mill had to be at least as high as its sails were long, while a merry-go-round could be lower, thus saving on building costs.[55] Furthermore, the mechanism to transmit power from the arms to the millstones or paddle wheels was somewhat simpler in a horizontally turning mill than in a vertical model, producing less friction. Finally, the merry-go-round would never have to be turned into the wind: it always faced in the right direction. Yet the Collegium also noted disadvantages. If, as proponents argued, the sails turned much faster on horizontal mills, their operation would be considerably more difficult. It might be possible to install a cover with lockable windows, as Mr. Pieter proposed and Stampioen had seen at Harlingen in Friesland, but then the advantage of not having to shift into the wind would be lost.[56] Another disadvantage was that the horizontal mill at Rijswijk had arms that were shorter than those of ordinary mills. Thus, the total surface area of sail was smaller, which was already the case because only one of the four arms stood properly in the wind with horizontally turning mills.[57]

The members of the Collegium did not limit themselves to theoretical speculations: they decided to build an experimental model of a merry-go-round. On January 8, 1627, they met in the carpentry shop of Abraham Willemsz. in order to examine the parts of the model. The costs would be shared by all, with the exception of Mr. Pieter, who had agreed to forgo claims on any potential profits.[58] Beeckman advised to make a model of a traditional windmill as well as a model of a horizontal windmill instead of making comparisons based on the performance of an existing mill. Thus, at the next meeting members agreed to make not just one, but two "little doll mills."[59] Small mills that were used in gardens to scare away birds would provide the necessary materials. Whether they actually built these models and carried out the proposed experiment is not clear. But Beeckman did use the principle of the merry-go-round in laying out the garden for burgomaster Puyck. Because the small mill in the garden was used only to produce a constant current in the water, its limited power was less important than the fact that the burgomaster did not have to shift the windmill into the wind.[60]

The operation of the merry-go-round did not lead to profitable results, but the group's experiences with another project—an attempt to maintain the depth of Rotterdam harbor—which was conducted simultaneously, were outright disappointing. For many years, traffic on the Maas was impeded by a sandbar that had formed directly in front of the harbor jetties on the northern shore of the river. Beginning in the last decades of the sixteenth century, the city government constantly made plans to solve the problem. Most of these projects involved building a dam on the Zuidergat, the arm of the river between the southern bank of the Maas and the island of Feyenoort in the middle of the river. The designers hoped to divert most of the river water to flow along the north side of the island, so that the sandbar that lay north of Feyenoort would be washed away. The government had such a breakwater built in the spring of 1626, which did not entirely close off the Zuidergat. But the breakwater project failed. Thus, on November 6, 1626, the Collegium Mechanicum proposed to consider "how one might be able to get the sandbar away from the jetty."[61]

The group discussed options. Drawing on the knowledge he had gained from studying Stevin's manuscripts,[62] Beeckman proposed moving the sandbar closer toward the southern bank of the Maas by placing ships in the river at appropriate points. At flood tide, they would be placed lengthwise, parallel to the banks, in order to allow free passage of the rising water. But at ebb tide, as the water flowed away, they would be placed sideways so that the water flowed over the sandbar. Obviously, the current would then remove some of the sand, and over the years the sandbar would probably disappear.[63] Beeckman also proposed building a low dam, which would appear above the surface only at low tide, between the south bank of the Maas and Feyenoort Island. Such a dam would resist the undercurrent, thus providing for the accretion of land to Feyenoort and the diversion of the current toward the north side of the island. The low dam would cost much less than a high dam that would close off the Zuidergat at high tide.[64] Mr. Pieter had a completely different plan: he proposed installing a floating platform with a merry-go-round mill above the sandbar. A wheel to stir up and loosen sand from the sandbar under the raft would be attached to the axle of the mill, which would never have to be shifted into the wind. The current would then carry away the loose sand.[65]

The members of the Collegium knew that the city administration was again holding consultations about the problem. On November 27, 1626, Beeckman and Stampioen were selected to represent the Collegium in consultations with the administration and to offer advice.[66] They learned of a plan to build a new dam in the Maas from Feyenoort Island before they were able to meet with the authorities. Therefore, they decided first to inspect the proposed site in a small boat and to make a map of the surrounding area, "and also to see if that jetty would serve the purpose, or not."[67] Despite the fact that Stampioen was an acknowledged cartographer and had already made a map of the Maas at Rotterdam in 1624, the city administrators decided to build the dam without consulting them. In January 1627, in the aftermath of their decision, Beeckman still offered to advise the magistrates privately to first make a good map of the site. Beeckman asserted it would have been "worthwhile if the magistrates asked our advice."[68] But that did not happen. When the city council decided on January 4 to build a new dam, one of the burgomasters was advised to seek counsel among "contractors and other persons having expertise,"[69] which did not include the Collegium. Therefore, Beeckman and Stampioen were relieved of their task at the Collegium's January 22, 1627, meeting. Its members then discussed the improvement of chimney flues.[70]

Such negative experiences strengthened Beeckman's initial doubts about the design of the Collegium. Members discussed only subjects they found interesting, whereas Beeckman would have preferred that the chairman have the right to decide the agenda.[71] He also noticed that the contributions of other members were not particularly interesting and that, as a result, his own influence on the Collegium was excessive. He could keep his opinions to himself until after the others had spoken, but this was not satisfactory either.[72] The Collegium depended too heavily on his contribution to maintain itself without him, which proved to be the case when Beeckman seriously considered leaving Rotterdam. On February 19, 1627, Beeckman was in nearby Dordrecht to discuss an offer to become principal of the Latin school there.[73] As soon as other Collegium members heard this news, they tried to give their society a stronger foundation, but to no avail.[74] Even before Beeckman left for Dordrecht in May 1627, the Collegium Mechanicum had collapsed.

PROBLEMS IN THE REFORMED CHURCH

The regents of Dordrecht were not the first to have asked Isaac Beeckman to be the principal of their Latin school. In May 1625 the magistrates of Den Briel, a city west of Rotterdam, had already asked him to serve as principal of their school. Despite the financial advantages, Beeckman had declined the offer.[75] Why then did he accept the bid from Dordrecht less than two years later? Could his decision have had something to do with the greater prestige attached to the Latin school in the stately city of Dordrecht? Or had things changed in Rotterdam between 1625 and 1627, making it less attractive for him to stay? The latter was indeed the case: problems in the Reformed Church had so poisoned the atmosphere for Beeckman in Rotterdam that in 1627 even a generous counteroffer made by the Rotterdam burgomasters—a higher salary and promotion to the principalship in case of the death or departure of his brother Jacob—did not change Isaac's mind about departing for Dordrecht.[76]

The persecution of Remonstrants, who had been expelled from the Reformed Church by the Synod of Dordrecht, underlay the religious problems. One could try to bring the Remonstrants back to the Reformed Church by means of a harsh persecution, or one might convince them to return voluntarily by clear preaching and the avoidance of unnecessary tests of conscience. Most authorities in Holland adopted the policy of severe punishment of the most brazen Remonstrants and simultaneous moderation toward well-intentioned Remonstrants, who were to be led back to the Reformed Church with a gentle hand. This policy was designed to achieve a broad public church instead of a small but strict Calvinist church. Most clergymen disagreed with this lenient policy but nonetheless accepted it as long as they could maintain their privilege of denouncing the lax regents from the pulpit from time to time.[77]

The fact that the Counter-Remonstrants constituted a minority in Rotterdam complicated the situation considerably. Rotterdam had long been a bulwark of the Remonstrants, which did not change after 1618. Of the thirty-five hundred members of the Reformed Church, some twenty-five hundred sided with the Remonstrants. Harsh persecution seemed unlikely to succeed in getting Remonstrants to return to the Reformed Church there; yet it was in Rotterdam that the Reformed clergy were particularly severe in its approach to the regents and members of the church. It did not accept the moderate policy of the local authorities and constantly pressured the burgomasters to take a harsher line.[78] Under the leadership of Van Berckel, chief of the moderate faction, the city council was unwilling to yield to the preachers and, in the course of the 1620s, showed itself very lax toward the Remonstrants. On many occasions, whether it was erecting a statue of the renowned humanist Desiderius Erasmus or granting a pension to the widow of a preacher, the city authorities and the preachers were diametrically opposed. The preacher Jacob van Leeuwen was particularly critical of the regents.[79]

At first Beeckman was not directly involved in these controversies, but that changed on June 8, 1625, when he was chosen as an elder of the Reformed Church. Within a month he was implicated in a conflict that escalated quickly. When Rev. Peter Moses died, a new preacher had to be appointed, and the city council had settled on Cornelis Hanecop, a clergyman from Breda, the town that had just been captured by the Spaniards. Hanecop was orthodox in doctrine but moderate in his practice toward Remonstrants, exactly what the regents of Rotterdam wanted. Although they did not want to impose their will on the church council, the city authorities in mid-June expressly suggested that Hanecop be appointed. A majority of the church council agreed, but a minority consisting of the preachers and a single elder refused to cooperate or state their objections to Hanecop's appointment. The majority, including Beeckman and his friend Weymans (an elder since 1621), therefore presented the matter to the classis of Schieland, which discussed it on July 1, 1625.[80] The classis then recommended that the offer to Hanecop not be made because his appointment would be divisive.

The majority was not satisfied with this outcome. Although in the meantime Hanecop had accepted a call to Amsterdam, thus removing the direct cause of contention, some procedural questions needed to be investigated.[81] At an extraordinary meeting of the classis on July 15, Beeckman, Weymans, and a third elder, Andries van Goch (who also was a member of the city council), presented two questions.[82] First, should not the preachers have expressed their objections to Hanecop in public? Second, could the preachers simply ignore a majority decision of the church council? Initially, there was no agreement on these points; then a session of the regional synod of South Holland worked out a solution. On September 26, 1625, a special commission produced an act of reconciliation, which was signed by all the clergy and nearly all of the elders, including Beeckman.

One elder—Weymans—refused to sign. The whole matter then was reopened because, while the commission of reconciliation wanted to take disciplinary measures against Weymans, most elders in Rotterdam, including Beeckman, did not. Appearing before the classis on March 12, 1626, Beeckman and his colleagues declared that the actual problem had not been resolved, despite the appearance of reconciliation. Did the clergy have the right to reject a candidate chosen by a majority of the church council? They believed that it was unjust to shift the discussion to Weymans's refusal to sign. The classis then called a new meeting for March 30, when several delegates from the provincial synod would also be in attendance.

In the meantime, the affair had escalated further: Weymans tried to prevent other members of the Reformed community from taking part in the Lord's Supper, and he and his supporters also had brought the matter before the courts. Furthermore, the clergy protested "that these malcontent elders are holding conventicles in the school, making leagues and alliances among themselves, and sticking so close to each other that they say, 'If you [try to] get one of us, you will have to deal with the others too.' "[83] In other words, the situation began to resemble a schism in the church, and the Latin school served as the headquarters for the dissidents.[84] As principal, Jacob Beeckman must have played an important role behind the scenes. He had no official function in the church, and his name does not appear in the documents. Yet Isaac Beeckman must have carefully discussed everything with his brother. Jacob also may have had a hand in writing the act of resistance that the dissident elders presented at the March 30 meeting, because these elders declared that "such was not their own work, but that they had used the help of others."[85]

Although Isaac Beeckman did not sympathize with the Remonstrants' beliefs, he had good reason for taking a strongly partisan role in this controversy. Beeckman's doctrinal orthodoxy was as strong as that of the clergy with whom he disagreed. He rejected the religious and political views of the Remonstrants.[86] Calvin was the most important theologian for Beeckman; and if the city regulations had not prevented it, he would have taught from Calvin and the Heidelberg Catechism instead of the pagan authors at the Latin school.[87] In the matter of Sunday observance, he took the position that the precepts imposed on the Jews for the Sabbath were still fully in force for Christians; he thus belonged to the more orthodox wing of the Reformed Church. Beeckman nevertheless maintained friendly relations with Remonstrants such as Stampioen and wanted to persuade them by preaching and setting a good example, rather than by persecuting them. Beeckman's views were pietistic, while his character was essentially peace loving. He was, as described above, a follower of Willem Teellinck, the pietist and mystically inclined preacher from Middelburg. Finally, Isaac and Jacob Beeckman did not yield to the preachers because they themselves had been educated for the clergy and did not consider themselves inferior to the Rotterdam ministers. In their own way, Jacob and Isaac Beeckman had a responsibility for the cure of souls both inside and outside the walls of the school. Isaac Beeckman had become the head of his family after the death of his father on December 2, 1625, and he supervised the doctrinal purity of his family members.[88] In 1623, when the sheriff of Zevenhuizen, Claes Bontebal, was condemned to death for his part in an unsuccessful plot to murder Prince Maurice, Jacob Beeckman had been one of those who attended Bontebal in the last hours before his execution.[89] The two brothers followed their father's example: refusing to heed the dictates of the preachers of Rotterdam as Abraham Beeckman had fought the dictates of the preachers of Middelburg.

Finally, on March 30, 1626, the day of the meeting of the classis, the city council of Rotterdam, led by Van Berckel, produced a new "proposal of moderation" that sought to reconcile the parties.[90] The dissident elders agreed to this proposal: Beeckman was after all a good friend of Van Berckel, "my truest friend that I had in Holland," as he wrote after Van Berckel's death in 1634.[91] The preachers, on the other hand, rejected the proposal and received the support of the synod of South Holland. Nonetheless, on July 10, 1626, an agreement was reached on a new selection process for clergy, replacement of part of the church council, appointment of two new preachers, and temporary employment of several preachers from outside the city.[92] When the agreement was read from the pulpits on July 12, the conflict appeared to have ended.

A return to normal, however, was impossible. The atmosphere in Rotterdam remained tense, and new disputes between moderate and orthodox factions on the church council and in the city government were bound to arise. The moderate Van Berckel remained in opposition to the harsh Van Leeuwen and his colleagues, some of whom were also regents of the Latin school and in this capacity could make life difficult for Beeckman. Isaac Beeckman did not wait for new conflicts to arise. Just as his friend Weymans had left Rotterdam shortly after January 1627, Isaac seized the opportunity to move to Dordrecht that spring. The burgomasters offered him, as stated above, an increase in salary and prospects of promotion to principal in an attempt to get him to stay. But Beeckman parted ways with the Rotterdam burgomasters on good terms: "They gave me an honest testimonial," Beeckman noted, "and honored me with a silver platter, as a sign that my service here was pleasing to them."[93]

CHAPTER THREE

Among Patricians and Philosophers, 1627–1637

Dordrecht, as the oldest city in the province, enjoyed many privileges in the assembly of the States of Holland. The city had the right to appoint the state pensionary—the highest administrator of the States—who also acted as the leader of the delegation of the States of Holland to the States General, the general assembly of the Dutch Republic. As such, the pensionary sometimes functioned as the informal leader of the Dutch Republic, especially during those years when the province of Holland had no stadholder. Yet Dordrecht's economic position had long ceased to correspond to its official status. The staple right, which had favored the city economically, had eroded in the course of the fifteenth and sixteenth centuries. Specialization in sectors such as the wine and wood trades and the salt industry offered some compensation, but not enough. Other cities, especially Amsterdam, and around 1620 even Rotterdam, its old rival, surpassed Dordrecht. Enterprising merchants began to leave the city in order to carry on their business elsewhere, despite the fact that Dordrecht, with a population of more than eighteen thousand, remained an important city.[1]

The stagnant economic life greatly influenced the choices and lifestyles of the regents of Dordrecht. Their active orientation as entrepreneurs gradually gave way to a more passive role as rentiers. Dordrecht regents no longer took an active personal role in trade; rather they invested their capital in ventures organized by others, something that was equally profitable. The regents increasingly devoted themselves to political matters.[2] After 1651, for example, the powerful Dordrecht regent Jacob de Witt gradually withdrew from the lumber trade and focused entirely on provincial and city governmental issues. A few regents exclusively occupied higher administrative offices. They took steps to hold newcomers and other wealthy merchants at bay so they could maintain their hold on city government.[3] These regents increasingly distanced themselves from ordinary citizens and adopted an aristocratic lifestyle, purchasing manorial rights and building beautiful country houses.[4]

Dordrecht thus began to acquire a refined character during the time that Beeckman lived there. Seventeenth-century historians praised the good manners of the people of Dordrecht, comparing them favorably with the inhabitants of Brabant, who set the standard for cultural refinement.[5] In 1662 one such historian reported that Dordrecht's inhabitants were more enamored of the arts and sciences than most Dutchmen. Dordrecht's elites, as it was said, studied even the most difficult subjects, including those that were perhaps beyond human understanding. Poets and writers settled in Dordrecht.[6] Dordrecht thus acquired a reputation for the artistry and learned nature of its inhabitants.

Thus, when the rector of the Dordrecht Latin school died in October 1626, the regents sought a successor who had a certain style. The regents consulted some professors at the University of Leiden, who in turn recommended Isaac Beeckman. Beeckman had acquired a reputation as a good teacher as well as an accomplished mathematician and natural philosopher.[7] On January 27, 1627, several Dordrecht regents gauged Beeckman's interest in a possible appointment; on February 15, he accepted their offer; and on February 20, the appointment officially was confirmed. The regents fixed Beeckman's salary at six hundred guilders, plus a portion of the tuition money; they also exempted him from excise taxes, and gave him free housing and heating fuel. He also received three hundred guilders toward moving expenses and two hundred guilders to hire several "tutors" (hypodidasculi) to live with the boarding students and help them with their homework.[8]

TEACHING AND LEARNING IN DORDRECHT

Beeckman faced a challenging workload in Dordrecht. His responsibilities included teaching the highest class, private teaching, and caring for about sixty boarding pupils. In addition, he had to maintain all the school buildings.[9] He took his duties seriously: for example, when he installed a new lock himself "to the great applause of the craftsmen, since it was an invention in an old craft by someone not schooled in it."[10] Naturally, others helped: his wife Cateline cared for the boarders; a vice-rector, two Latin masters, and one "Dutch" master also assisted him.[11] Yet Beeckman struggled to keep up with the demands of his job. He complained repeatedly about his "very heavy burden, exceedingly inconvenient for every study."[12] The "board of governors" (scholarchi), who supervised the school for the city, soon began receiving complaints regarding the lack of order and regularity in the functioning of the school. For example, some complained about the numerous holidays granted by the rector as well as the irregular starting times of classes. New regulations helped to some extent: Beeckman abolished some holidays, had a bell installed to mark the start of classes, and in 1635 began closely supervising a particularly lax teacher.[13] In principle, Beeckman approved of stricter rule, but not in practice, devoting too much time to his own studies to keep order and discipline at the school.[14] Nonetheless, the school flourished under his leadership as never before. Beeckman was very popular with his pupils, and the number of students grew accordingly, so that he had to hire additional staff.[15] On December 1, 1629, Beeckman's younger brother Abraham was appointed Latin master, undoubtedly at Beeckman's behest.[16]

Isaac Beeckman formally assumed office on June 2, 1627, and delivered an inaugural lecture on "physical-mathematical philosophy," probably in one of the churches of the city. Beeckman showed how the knowledge of isoperimetric figures—mathematical figures with equal perimeters, but different surface areas—could explain many natural phenomena. This was a rather unusual subject for a rector of a Latin school, and the public undoubtedly felt more at ease with the much more traditional oration by the preacher Balthasar Lydius, which had preceded Beeckman's lecture. (Lydius served on the board of governors.)[17] Still, it is a very remarkable oration. Beeckman tried, in his own way, to present himself as a man of polite interests and erudite learning, but this learning, so he argued, included the study of what he called physical-mathematical philosophy. If the magistrates provided him with the necessary support, Beeckman asserted that the introduction of this philosophy could restore "the learned golden age of antiquity" (saeculum doctum et vere aureum).[18] It is instructive to compare Beeckman's oration with the much more famous lecture delivered by the humanist and philologist Caspar Barlaeus in 1632 in Amsterdam, on the occasion of the inauguration of the Amsterdam Athenaeum illustre. In this lecture, entitled "Mercator sapiens," Barlaeus exhorted the Amsterdam merchants, who financed the new institution of higher education, to study philosophy and mathematics, just as Beeckman did in Dordrecht. But whereas Beeckman presented his audience with a new kind of philosophy, Barlaeus referred only to ancient philosophers and mathematicians, never mentioning modern developments in these disciplines.[19]

The regents of Dordrecht did indeed support Beeckman's studies: in September 1628, soon after his arrival, the city paid for a small tower built on the Latin school's roof (the school was located in an old convent). The tower was specially designed for meteorological and astronomical observations.[20] This tower—probably no more than a raised platform roof on the chapel of the convent—was the first official observatory in the Dutch Republic. (The first *university* observatory was established in Leiden in 1632, four years later.) Thus, one of Beeckman's most deep-seated wishes was fulfilled. Beeckman had tried to make systematic meteorological observations when he lived in Zierikzee, but the series he then recorded were irregular and highly inaccurate. In Zierikzee, he had to check the direction of the wind by looking at the weather vane on the tower of the church, whereas in Dordrecht he could tie the weather vane on his tower to an indicator in the room beneath it and thus precisely determine the wind direction.[21]

Beeckman kept other instruments in the space below the platform roof, which he referred to as his "musaeum."[22] In June 1630 he observed a solar eclipse with a "tube in a dark space" (per tubum in loco obscuro), which was a cylinder with a pinhole aperture through which the light of the sun and the shadow of the moon were projected on a piece of paper in the darkened musaeum.[23] The eclipse reached its maximum that day at seven degrees above the horizon, so Beeckman must have projected the image of the sun on the wall of his musaeum. A year later Beeckman also acquired a telescope from a lens grinder in Delft. The people of Dordrecht attributed magical powers to the telescope: people believed, as Beeckman notes in the *Journal*, that he had seen a ship, supposedly lost, unloading cargo at some Brazilian port through his telescope.[24]

More interesting even than the telescope was another instrument, the thermoscope, which indicated the air temperature variations. The thermoscope consisted of a long glass tube, closed at the top, with a glass bowl; its base stood in a container of water. When it was warm, the air in the glass bowl expanded and the water level in the tube fell; the reverse occurred in cold weather.[25] According to Beeckman, the instrument, which he referred to as his "instrumentum Drebellianum," had been invented by the Dutch alchemist and engineer Cornelis Drebbel, who at that time lived in England.[26] Beeckman's thermoscope was unusual: the glass bowl was on the roof, exposed to the open air, while the glass tube ran through the roof to his musaeum below. Thus protected from the elements, he could record the exact height of the water column.[27]

All through his life Beeckman was fascinated by the figure of Cornelis Drebbel (1572–1633). Drebbel was the son of a farmer in the neighborhood of the city of Alkmaar, became an engraver and an engineer, and in 1605 entered the service of the English king James I. He was known for inventions like the submarine and the microscope, but his main claim to fame were his perpetuum mobile constructions, including his thermoscope. During a visit to Middelburg in November 1619, Beeckman was able to consult Drebbel's *Treatise on the Nature of the Elements* (*Tractaet van de Natuere der Elementen*), a short exposition of Drebbel's alchemical and mystic philosophy of nature, published in 1604.[28] This first edition had only a small print run, and Drebbel had sent it to "good friends and to philosophers" only. The fact that Beeckman was able to consult this rare edition (no copy of it survived) indicates that either his father or Philip van Lansbergen belonged to this inner circle of Drebbel's friends. Beeckman, as one can see from other notes in the *Journal*, always secretly admired Drebbel, even though this would-be philosopher and inventor was almost aggressively anti-academic and claimed that his mysterious instruments directly validated his theories about the nature of the world without the need for scientific reasoning. Through his father Beeckman was constantly informed about the whereabouts and achievements of Drebbel. On March 15, 1620, Beeckman received a letter from his father, informing him about the famous submarine, and in 1631 the transcription of a letter written by Drebbel to the English king in 1613 again testified to Beeckman's contacts with the circle around Drebbel.[29] The drawings that accompanied this letter show the first illustration of the compound microscope, the invention of which is also ascribed to Drebbel (see fig. 2).

Notwithstanding his admiration for Drebbel's inventiveness, Beeckman's interpretation of Drebbel's instruments was completely different from that of Drebbel himself.[30] For Drebbel, an instrument like the thermoscope was not merely an instrument for measuring heat but a perpetual motion machine that exemplified the hidden sympathies in the larger world. The rise and fall of the water in the thermoscope resonated with the cyclical movements in nature: it demonstrated the breathing of the cosmos (it was therefore also a cosmoscope). With these "living instruments" Drebbel stands out as a representative of a machine-based but nonmechanical philosophy of nature that remained in vogue far into the seventeenth century as a healthy reminder that the mechanical philosophy could still contain many vitalistic elements. This mechanization of the world picture was not some sudden "Gestalt switch" in natural philosophy but an ongoing battle between two related and yet completely different approaches of nature.[31] Now Beeckman clearly belonged to the opposite camp. In his comments on Drebbel's "living instruments," he completely stripped off their mystical connotations. For instance, instead of the vitalistic explanation

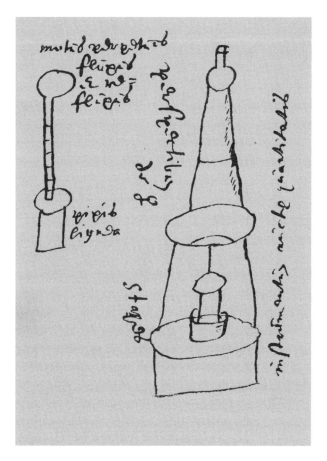

Figure 2. Beeckman's drawing of Drebbel's thermoscope and
microscope (*Journal*, 3:442)

of the self-regulatory character of Drebbel's "perpetual clock" (horologium
perpetuum), Beeckman speculated that the fact that the hands of the clock
threw a shadow on the sundial and therefore caused a local drop in temperature
could very well explain why the hands adjusted to the correct time during the
day.[32] In the same vein, Drebbel's thermoscope according to Beeckman was not
a mystical demonstration of the cyclical movements in the cosmos but a simple
instrument to measure heat, nothing less, nothing more. As he had done with
Aristotle's natural philosophy and Galen's physiology, Beeckman accepted the
substance of Drebbel's inventions but gave them a completely new, purely mechanical interpretation.[33]

Although the rooftop observatory was built especially for Beeckman, he did not make his observations alone. He regularly invited friends and dignitaries to study the skies from his tower and even employed Latin school students to make observations and take measurements. During the eclipse of 1630, students kept track of time, while Beeckman chronicled the progress of the shadow of the moon. Afterward, Beeckman allowed students to carry out orders independently and use his instruments to conduct experiments. Even the maid was sometimes asked to continue an experiment in the absence of the master.[34]

Beeckman thus had reason to be satisfied with the support he received from the regents; however, he viewed the regents' support for scientific work as inadequate. Shortly after taking office, Beeckman writes, "The gentlemen make such great expenditures on the Latin schools, whereof they seldom enjoy the fruits themselves; why should they not spend a bit to educate the citizens in natural sciences and mathematical arts, which would be useful to their city?"[35] Before his departure from Rotterdam, Beeckman had toyed with the idea of starting with a "philosophical course" that would not include theology or politics, as he explicitly stated.[36] Thus, one of Beeckman's pupils would present a theorem of Euclid every Saturday morning at ten in the Latin school's auditorium; Beeckman himself would then explain the theorem's practical application. Such lessons would have to be in Dutch because they were geared toward carpenters, masons, ship captains, and other citizens. A year later, just as he was reading Francis Bacon's *Sylva Sylvarum* (1621), Beeckman proposed the establishment of a "college" in which people who were experienced in the mechanical arts and others could meet to discuss new developments in the field of natural inquiry and the arts.[37] Beeckman wanted the magistrates to support the Collegium by paying attendance fees. The new project would have been the first scientific society in the Netherlands.[38] The regents were not interested, however, and nothing came of Beeckman's plans.

Beeckman soon felt completely at home in Dordrecht, although his ties to Rotterdam remained strong. His brother Jacob and sister Maria still lived there, as did his brother, Abraham, who had been hired as a teacher at the Latin school after Isaac's departure. After Beeckman's father's death in 1625, his mother also probably moved to Rotterdam to live with her sons. In 1629, however, things changed dramatically: Beeckman's mother died in June, and in August his brother Jacob died of consumption. Jacob's death must have represented a great loss for Isaac Beeckman, who always had relied heavily on his younger brother. Shortly afterward, when Abraham was employed at Dordrecht and Jacob's widow came to live there, Isaac no longer had much reason to travel to Rotterdam. Dordrecht became the new center for the family, and he, Isaac Beeckman, more than ever, became its titular head or *pater familias*.[39]

The common people of Dordrecht began spreading rumors that Beeckman could predict the future by gazing at the stars after watching him make observations from his tower at all hours of day and night. Thus, in February 1628, shortly after the lunar eclipse of January 20, some people believed that Beeckman had foreseen that three Dutch cities would fall to the hands of the Spaniards that year; in 1631, people spread another rumor that Beeckman had seen in the stars that on a particular day fire would break out in four places in Dordrecht.[40] When a fire started on the eve of the feared day, people called out in the streets, "Now it's happening, just like the rector predicted!". Yet Beeckman had no faith in astrology: he regarded the powers popularly attributed to him as a warning not to believe any similar stories told about others.[41]

Beeckman was fascinated, nonetheless, by the boasts of alchemists and astrologers and—if they did not first approach him—made a point of visiting them to learn the basis of their pretensions. One asserted that he could extract the "quintessence" of Rhine wine and make unlimited amounts of new wine. Another declared that he could heal a wound without touching it; yet another claimed to know something about Drebbel's perpetual motion machine. But the most extraordinary figure was undoubtedly a well-to-do miller of Gorkum, Balthasar van der Veen (or Van der Vinne), who held numerous remarkable cosmological views. For example, he maintained that the Earth was hollow and that people were living inside it; he also believed that at an altitude of two miles, the air thickened to a crystal sphere, pieces of which broke off from time to time and could be observed as clouds or mist.[42] Van der Veen had not come up with such ideas on his own: he had picked up various ideas from Giordano Bruno and also believed in Copernicus's theories. He believed in both the annual and the daily rotation of the Earth, and showed that stellar parallax, which should be observable according to theory, was in fact observed. Indeed, in astronomical books, he sometimes had encountered different positions for the fixed stars.[43] Although Beeckman rejected most of Van der Veen's notions as "trifles" (nugae), the extraordinary ideas of the self-taught miller of Gorkum fascinated him.[44]

Representatives of traditional learning also lived at Dordrecht, and Beeckman regularly profited from their erudition. Beeckman had access to international and cutting-edge scholarship thanks to these new acquaintances. The Walloon preacher Andreas Colvius, who was called to a ministry in Dordrecht in 1629, was the most significant of these new friends. Colvius was renowned for his collection of books and manuscripts, and his richly supplied cabinet of natural history.[45] In 1622, when Colvius served as a preacher in the Dutch embassy to the Republic of Venice, he began to build his collection. He remained in northern Italy until 1627, establishing contacts with many Italian scholars, and he procured manuscripts of such leading lights as Paolo Sarpi and Galileo Galilei. Beeckman gratefully made use of Colvius's manuscripts, developing, for example, an extensive knowledge of Galileo's works.[46] Colvius, a "careful inquirer of natural things," also made astronomical observations with Beeckman in his little tower until just before the latter's death.

Through Colvius, Beeckman became acquainted with Jacob de Witt, one of the most powerful regents of Dordrecht. Although stiff in manners and strict in religious matters, De Witt was not a disagreeable man. During the 1620s and 1630s, he regularly received friends at his house beside the Great Church to discuss literature, theology, and other learned matters. Beeckman was among De Witt's regular guests, although his literary taste was out of sync with current fashion. Jacob Cats of Zeeland, who had been pensionary of Dordrecht since 1623, was the leading poet in De Witt's circle.[47] (Beeckman had met him in 1620 when he and Cats had both been present at the examination of Daniel Nota's new invention for dredging the port of Middelburg.) Although the edifying moral message of Cats's poetry certainly must have appealed to Beeckman's pietist sympathies, he disliked Cats's literary style. Beeckman was somewhat old-fashioned, as he was still moved by the relatively simple rhymes of the rhetoricians, whose works were meant to be read aloud.[48] Cats wanted above all to be a *poeta doctus*, a learned poet, and he wrote poetry to be read silently and studied. Besides literature, Cats and Beeckman may also have discussed the poet's involvement with some draining projects in Flanders and England. Other ties bound Beeckman to the De Witt circle: at the request of De Witt's wife, he repaired the pumps of the regent's house.[49] Beeckman never concealed his technical background, and people repeatedly sought his advice about technical matters.

Beeckman's contact with Cats was limited by the latter's frequent travels to The Hague, to represent Dordrecht's interests. Beeckman had much more frequent contacts with Dr. Johan van Beverwijck, who was a city physician, member of the city council, and a regent of the Latin school.[50] A polymath, Van Beverwijck's talents ranged from writing a history of Dordrecht to laudatory poetry. In 1634 he published a collection of essays on the physician's role in the light of divine predestination with contributions from luminaries such as the rector of the Latin school at Delft, Jacobus Crucius, the French monk and scholar Marin

Mersenne, and the famous humanist Caspar Barlaeus from Amsterdam. Andreas Colvius also contributed an essay to the volume. Learned conversations at the homes of De Witt and Cats centering on "the secrets of nature, necessity, the providence and foreknowledge of God, and the fixed term of our lives" gave rise to Van Beverwijck's collection.[51] William Harvey's ideas about blood circulation, published in 1628 in his *De motu cordis*, were discussed in this collection. After some initial hesitation, Beeckman accepted and propagated Harvey's theory at least from 1633 onward. When De Witt's wife Anna became seriously ill that year, Beeckman prescribed a treatment, which proved effective, on the basis of Harvey's theory.[52] As a result, others recognized the new theory's validity, including Van Beverwijck, who published a treatise on kidney stones in 1638 in which he spoke out in favor of Harvey's theory, thus becoming the first in the Dutch Republic to do so publicly.[53]

Thus, within a short time Beeckman had established his place in Dordrecht's intellectual and cultural life. Although he was shy, members of the elite respected his opinions and intelligence. He now moved freely in the highest social circles of a prestigious city like Dordrecht after having started his career as a simple candlemaker in Zierikzee in 1611. Eminent preachers and physicians served as godfathers to his children.[54] By the early 1630s, Beeckman had achieved success in life. The fact that in 1628 Beeckman gathered all his manuscripts together in one volume and bound this book in calf with copper mounting, more or less like a family bible, represents a clear sign of his self-confidence and his place as a learned man with polite manners.[55] He added marginal headings that made consultation easier and he also inserted documents that he had kept separate, such as Descartes's 1618 note on the derivation of the law of falling bodies. The notebooks, at first conceived as a traditional commonplace book, had finally been transformed into the *Journal*. Beeckman, however, held on to the custom of writing new ideas first in a scratch pad (in Dutch, "tafelboeckxken").[56] Afterward, he hired someone to copy them into his *Journal*.[57]

ENTERING THE REPUBLIC OF LETTERS

Besides the learned and literary elite of Dordrecht, Beeckman began to establish contacts with natural philosophers in foreign countries. The relative isolation in which Beeckman had developed his ideas came to an end around 1630, and he became a member of the international Republic of Letters. Beeckman received a completely unexpected visit from his old friend René Descartes on October 8, 1628.[58] They had not seen each other since their meeting at Breda in late 1618. In 1619 Descartes had begun an adventurous journey through Germany, and while there—so he later reported—he laid the basis for his philosophical system, the first glimpses of which would be published in his *Discourse on Method* in 1637.[59] After a year or two, he had returned to France and remained there except for a trip to Italy to settle some family business (between 1623 and 1625). In the social realm, Descartes had not achieved anything: he had not acquired a landed estate, nor had he bought a judicial office with a regular income, and he still had not married. Instead, he had sold most of his properties, receiving enough interest to live independently in modest circumstances. In Paris, he devoted his life to mathematics and natural philosophy. By the end of the 1620s he felt a growing pressure to publish some of his ideas. He was working for quite some time on his *Rules for the Right Conduct of the Mind* (the *Regulae*), which several people had urged him to finish. According to Descartes, Paris was not the right place to do so, because of the bad air and the distractions that accompany social life there. A retreat to the countryside in Brittany in the winter of 1627–28 did not help, because family members and neighbors distracted him. Therefore, in 1628 Descartes considered a more radical move: leaving France to escape from all social obligations. Italy was too hot, a war raged in Germany, but the Netherlands provided him with a reasonable alternative.[60] He had been to the Netherlands in 1618–19 and apparently liked it. So Descartes made an exploratory trip to the Dutch Republic in the fall of 1628 and, during this trip, also paid a visit to his old friend Isaac Beeckman. When Descartes did not find Beeckman in Zeeland, he was directed to Dordrecht; thus in October 1628, he appeared on Beeckman's doorstep.

The renewed friendship was very warm. Descartes told Beeckman about the progress he had made since 1619, about his discovery of the law of refraction of light, and about his new philosophical ideas. He proposed to Beeckman that they work together to finish his researches. Descartes claimed that, in all his travels, he had not found anyone to whom he felt as close as to Beeckman.[61] For his part, Beeckman offered Descartes his warmest praise. Mathematics again formed the principal subject of discussion, and Beeckman was especially intrigued by Descartes's findings in what would later be known as analytic geometry. When Descartes departed, he promised to send Beeckman a copy of his unpublished "Algebra" on his return to Paris. Indeed, Beeckman received this manuscript at the end of January 1629. In an accompanying letter, Descartes also approved a solution that Beeckman had found for a mathematical problem—a sort of brainteaser—that Descartes had left behind on his departure from Dordrecht.[62]

Descartes had returned to Dordrecht by the end of March 1629. He was on his way to Amsterdam and stopped briefly in Dordrecht to greet Beeckman.[63] Descartes traveled from Amsterdam to Franeker in Friesland, where he matriculated at the local university on April 16, 1629. A letter Descartes wrote to the French instrument maker Jean Ferrier in Paris evoked the intimacy of Descartes's relationship with Beeckman at that time. Descartes tried to persuade Ferrier to come to Franeker to work with him in perfecting hyperbolic lenses. Everyone else was working with spherical lenses, but while still in Paris Descartes had figured out that hyperbolic lenses would create a sharper image. Not being a craftsman, Descartes needed an experienced lens grinder to build a lens-grinding machine that would produce hyperbolic lenses.[64] In this letter, Descartes gave Ferrier advice on the route to follow and suggested a visit to Beeckman in Dordrecht. If Ferrier needed anything for the journey, Beeckman would undoubtedly give it to him on seeing Descartes's letter.[65] Beeckman would also give him the address where he could find Descartes (or forward him to someone who knew). However, Ferrier chose not to go.

The philosopher and canon Pierre Gassendi did travel from France to the Netherlands. As a philosopher, Gassendi had publicly proclaimed himself an outspoken anti-Aristotelian.[66] From 1617 to 1623, he had been professor at Aix-en-Provence and raised all possible objections in his lectures on Aristotelian philosophy. His book *Exercitationes paradoxicae adversus Aristoleos* (Paradoxical Exercises against the Aristotelians, 1624) was also primarily a skeptical attack on the prevailing philosophy. Yet over time, he had come to the conclusion that it would be better to build an alternative to Aristotelianism rather than taking an exclusively skeptical stand. For that alternative, Gassendi had in mind the equally ancient philosophy of Epicurus. Although Epicurus had a bad name as an advocate of hedonism in Christian Europe, Gassendi maintained that there were possibilities for rehabilitating this despised philosopher. Gassendi thus was engaged actively in the study of Epicurus when he traveled through the Dutch Republic with the nobleman François Luillier in 1629. As a practicing libertine, Lullier was the opposite of the sober and devout Gassendi, but he was willing to pay for the trip.

Gassendi first visited the southern Netherlands, but on July 2, 1629, he arrived at the harbor of Vlissingen in Zeeland.[67] He traveled via Middelburg to The Hague, Leiden, Amsterdam, Utrecht, and Rotterdam, then back to Leiden. On July 14, he headed to 's-Hertogenbosch in Brabant, where the army of stadholder Frederick Henry was besieging the city. At that time, sieges were a kind of tourist attraction because of the ground-breaking technology involved.

Noblemen and engineers, as well as philosophers such as Gassendi, traveled to witness such spectacles and to meet interesting people.[68] On his way to the siege, Gassendi passed through Dordrecht and visited Beeckman. This visit did not last long and was probably limited to a simple exchange of ideas. Beeckman gave Gassendi a detailed summary of his natural philosophy, paying special attention to his principle of inertia and his atomistic theory of matter.[69] Gassendi recounted his plans to rehabilitate the philosophy of Epicurus and specifically his ethics. The two exchanged gifts: Gassendi gave Beeckman a copy of his *Exercitationes paradoxicae* as well as a report on the curious meteorological phenomenon called the parhelia. In March 1629 the astronomer Christopher Scheiner in Rome had observed these "false" or "multiple suns." Gassendi had acquired several copies of Scheiner's report—which he distributed among his friends—through his patron Nicolas-Claude Fabri de Peiresc and now presented one to his Dutch host.[70] Beeckman, in turn, gave Gassendi a copy of the corollaries of his dissertation.[71] On his return from the siege, Gassendi paid a second visit to Beeckman and then returned to Paris.[72] In a letter to Peiresc, written in Brussels on July 27, 1629, Gassendi explicitly called Beeckman "the best philosopher I have met thus far" (le meilleur philosophe que j'aye encore rencontré).[73] The fact that someone like Gassendi viewed Beeckman this way means that he must have been excited about what he had heard from his host. Gassendi was back in Paris on August 8.

Although Gassendi's contact with Beeckman had been limited to an exchange of ideas and writings, the visit had a considerable impact on the development of Gassendi's thought. Gassendi had listened to Beeckman very attentively and must have been glad to meet another thinker who was not consumed by Aristotelian prejudices: Beeckman combined his atomism with a belief in God's creation of the world, and thus his thought was free of the atheism, which, according to many observers, was inherent in atomism. Perhaps he observed as well that Beeckman, although Protestant, was a devout and sincere Christian: Beeckman did not simply pay lip service to Christian beliefs. Therefore, during his return journey to Paris, Gassendi decided that his rehabilitation of the philosophy of Epicurus would include not only his ethics but also his atomistic philosophy of nature. Beeckman subsequently professed surprise at this decision, because during their conversation Gassendi had spoken only about Epicurean ethics.[74]

Meanwhile, Beeckman had established contacts with a third important French scholar and natural philosopher, the Parisian friar Marin Mersenne. When Descartes visited his friend in Dordrecht in October 1628, he had told him about a "monk" in Paris who had given him some information about the thickness of a string of a musical instrument and its pitch.[75] Beeckman then asked the Leiden professor of theology André Rivet about this friar's identity. Rivet was a Huguenot who had emigrated from France to the Dutch Republic in 1620, yet maintained close contacts with French scholars, often acting as an intermediary between French and Dutch scholars.[76] (Two of Rivet's sons had studied under the Beeckman brothers at the Latin school in Rotterdam.) Now, Rivet informed Beeckman that the friar was Marin Mersenne, whom he had recently contacted in connection with theological questions.[77] Rivet then wrote Mersenne advising him of Beeckman's interest in corresponding with him about questions in the field of musical theory and acoustics. As Paris was celebrating the fall of the Huguenot stronghold of La Rochelle, Mersenne replied that he welcomed such a correspondent.[78] Rivet sent Beeckman Mersenne's *Synopsis mathematica* and Beeckman received his first letter from Mersenne in early March 1629.[79]

The Jesuits of La Flèche taught Marin Mersenne, just like Descartes, but—in contrast to Descartes—he had gone on to study theology in Paris.[80] In 1611 Mersenne entered the Order of Minims, a branch of the Franciscans, and his superiors in the order assigned him primarily apologetic and scholarly tasks. Beginning in 1623, he published a series of scholarly studies in defense of orthodox Catholic religion. Furthermore, from his monastery off Place Royale in Paris, Mersenne maintained such a vast correspondence with scholars that he eventually acquired the title of "general secretary of learned Europe."[81] Writing a letter to Mersenne was akin to publishing an article in a scientific journal (the first scientific journal, the *Philosophical Transactions*, which appeared from 1665 onward, actually grew out of the private correspondence of the secretary of the Royal Society, Henry Oldenburg). Mersenne also managed Descartes's correspondence with scholars and scientists in France; letters destined for Descartes first had to be sent to Mersenne, one of the few people in France who knew Descartes's whereabouts. Gassendi was also one of Mersenne's best friends. His enthusiastic reports of his encounters in the Netherlands inspired Mersenne's desire to travel there too. Mersenne, however, could not travel so freely, because he was a member of a regular religious order. So, Mersenne headed north in the spring of 1630 under the pretext that he had to take a water cure at the city of Spa in the southern Netherlands. He did indeed visit Spa but stayed such a short time that it was clearly not his principal aim.[82]

Mersenne left Paris on April 26, 1630 (or soon after), and he crossed the border between the Spanish Netherlands and the Dutch Republic at Sluis in the

Dutch part of Flanders. Mersenne had to take off his brown frock at Sluis because Roman Catholic clergy were not allowed to appear openly in the republic.
He traveled to Leiden, where he met Rivet, and in early August he visited Beeckman in Dordrecht.[83] Mersenne spent days reading Beeckman's *Journal*. The
two scholars held extended discussions about the theory of music, a topic quite
different from those Beeckman had discussed during Gassendi's visit. Mersenne had his own ideas about many subjects and (for example) was still too
much of an Aristotelian to accept Beeckman's principle of inertia.[84] After his
visit with Beeckman, Mersenne returned to the southern Netherlands, where
he briefly stopped in Spa. By October 15, 1630, Mersenne had returned to Paris,
continuing his correspondence and exchange of ideas with Beeckman, using
Rivet and the courier services of the French ambassador in The Hague as
intermediaries.

THE BREAK WITH DESCARTES

Beeckman enjoyed his new French contacts, especially the correspondence
with Mersenne. From Mersenne, Beeckman learned of the latest developments
in the intellectual world, while also engaging an expert in mathematics and
natural philosophy in discussions of his own ideas. Yet Beeckman's contact
with Mersenne led—quite unintentionally—to a break with Descartes. Beeckman had prided himself on his good relations with Descartes, and he was troubled when these came to an end in the 1630s. Beeckman's broken relationship
with Descartes cast a shadow over the final years of Beeckman's life.[85]

Mersenne had included several questions, probably about musical theory, in
his October 1628 letter to Rivet when he agreed to establish contact with Beeckman. Mersenne instructed Rivet to forward these questions to Beeckman and
to request a solution, thus establishing an exchange of ideas.[86] Yet Beeckman
was already familiar with these problems, because he had discussed them with
Descartes in 1618.[87] Beeckman, through Rivet, pointed this out to Mersenne
and said he assumed that Descartes had given these problems to Mersenne
without telling him that he, Beeckman, had already solved them. Beeckman repeated this point in a letter sent directly to Mersenne. He praised Descartes and
wrote, referring to one of the questions: "He is the same one to whom more than
ten years ago I communicated [communicavi] the things that I have written
about the causes of musical harmonies, and who, I assume, gave you the idea for
this question."[88]

Now this is a correct statement, as the *Journal* entries confirm. Yet even a relatively harmless statement could lead to big trouble when communicated to Mersenne. As Descartes's biographer Richard Watson writes, Mersenne "loved to get people into disputes, and by controlling the information people sent him, Mersenne could instigate and prolong arguments between almost any of them at will."[89] Mersenne was very talented at fomenting trouble, especially with Descartes, who had, as Mersenne knew very well, a very short fuse. So sometime in September 1629 (the letter is lost), Mersenne wrote to Descartes—then still in Franeker—about Beeckman's claim, exaggerating a bit to stir up some trouble, claiming that Beeckman bragged that he had been Descartes's teacher. Descartes was furious. In a letter to Mersenne of October 8, exactly one year after renewing his friendship with Beeckman, he thanked Mersenne for informing him about his Dutch friend's ingratitude. Descartes insinuated that Beeckman must have been flattered by the honor of corresponding with Mersenne. Furthermore, Descartes claimed Beeckman must have believed that Mersenne might think even more highly of him by claiming to have been Descartes's teacher: "qu'il a esté mon maistre il y a dix ans."[90] Descartes, who by now had moved from Franeker to Amsterdam, stopped corresponding with Beeckman, without giving any reason. He did, however, ask Beeckman to return his copy of the *Compendium musicae*, the New Year's present he had sent him on January 1, 1619.[91] In subsequent letters to Mersenne, Descartes assumed an arrogant and mocking attitude toward Beeckman. For example, in a letter to Mersenne from January 1630 Descartes claimed that Beeckman's theory regarding the material nature of sound was "ridiculous."[92] He also accused Beeckman of passing off ideas he had borrowed from others as his own, while in fact it was just the other way around. For example, Descartes stated that Beeckman had taken the principle of inertia from him when they derived the law of falling bodies in 1618, instead of the other way round.[93] While Beeckman still wrote of Descartes as "our friend" in his letters to Mersenne, Descartes's letters to Mersenne insinuated that Beeckman had taken the little that he knew from others. The fact that Beeckman had given Gassendi a copy of his *Theses* during Gassendi's visit to Dordrecht, as Mersenne had learned from Gassendi, served as proof to Descartes "that [Beeckman] has done nothing better since" (qu'il n'a rien fait depuis qui soit meilleur).[94] And when Mersenne asked Descartes for clarification of some points in Beeckman's notebooks that he had not quite understood, Descartes answered: "It is easy to see that the obscurity of his words hides nothing that we should regret not understanding" (Il est aisé a juger que l'obscurité de ses paroles ne cache rien que nous devions avoir regret de ne pas entendre).[95]

However, Descartes could not brush off the matter so easily. Mersenne had deemed it unwise to inform his Dutch correspondent about Descartes's anger before he visited Beeckman. Yet, when he had consulted Beeckman's *Journal* during his visit to Dordrecht and read about ideas Descartes had claimed were his in August 1630,[96] Mersenne could no longer refrain from telling Beeckman. Obviously, Beeckman was upset: he had always been conscientious about citing his sources, and now someone he regarded as a friend had taken advantage of him. He wrote to Descartes for an explanation and also asked Mersenne for clarification.[97] At first, Descartes did not respond. And when he reacted—after receiving a second request for an explanation—he did so rather condescendingly. In a letter written in September or October 1630, Descartes advised Beeckman that he should not have taken his 1618–19 compliments too literally because, as he said rather disingenuously, French custom prompted him to be unduly generous in that regard. Furthermore, Descartes claimed that Beeckman had learned more from him than he had learned from Beeckman. Even if Beeckman had letters proving that Descartes had copied something from him, he advised Beeckman against making a grand show of it, because Descartes's friends knew that he was willing to learn even from worms and ants.[98]

Beeckman was not put off by such arguments. In an early October, 1630 letter to Descartes, Beeckman summarized the ideas Descartes had taken from him.[99] When Descartes read this letter, he exploded and hit back hard in a long and extremely nasty letter of October 17, 1630. He indignantly rejected Beeckman's soothing words and his proposals for renewed cooperation. Beeckman's friendly and familiar tone angered Descartes, who interpreted it "as if you were writing to one of your pupils."[100] Furthermore, Descartes stated that the problems Beeckman considered to be worthy of further study were trivial, at most interesting to a beginning student. Because Descartes understood that Beeckman had acted more out of some sort of (mental) illness—"ex morbo"—than evil intention, however, Descartes gave him a few points to consider, so that he would not make the same mistake again.[101]

Descartes first denied having learned anything of value from Beeckman's "mathematico-physica." There are indeed, he explained, things that can be taught by someone else, such as languages, history, experiments, and geometric demonstrations. If, however, the matter being taught requires convincing reasons or reliable authority, properly speaking nothing is taught when these are missing. According to Descartes, this is precisely the case with Beeckman's philosophy. The things Beeckman thinks Descartes learned from his "philosophia physico-mathematica" were advanced without convincing reasons or reliable authority. Descartes's approval of Beeckman's ideas thus did not mean that he had been convinced by Beeckman's reasoning but simply that he already thought out the reasons himself. So even if he sometimes endorses the things Beeckman claims as his findings, there is no reason to believe that he learned them from Beeckman.[102]

Descartes added that Beeckman's ideas could not have influenced him, because Beeckman's manuscript contained nothing of significance. According to Descartes, there are three kinds of discoveries. The truly great discoveries arise from the power of a great mind and can belong only to those who have actually made these kinds of discoveries. Because such discoveries are so great, however, the discoverer need not fear that others could steal his ideas. Descartes implied that the discoveries in Beeckman's *Journal* could not be of this first order, because Beeckman was acting in such a petty manner. Discoveries of the second kind are made by chance. Because anyone can be lucky, it is of course understandable that anyone who has made such a discovery anxiously would try to protect it from others. Finally, the third kind of discoveries are those of limited value, which are regarded as highly significant only by the discoverer, just as a fool believes that the box of little stones and pieces of colored glass found among his neighbors' rubbish constitute real treasure. "I do not wish to compare your manuscript to such a little box," Descartes added, "but I think that hardly anything more solid could be found in it than bits of gravel and glass."[103]

Descartes clearly intended to crush Beeckman psychologically in this letter, but why was Descartes so vicious toward him? Descartes had outgrown Beeckman in many ways: Descartes even reminded Beeckman how on renewing their friendship he had informed him of some of his new findings and that Beeckman had admitted that he did not understand them immediately. By the 1620s, Descartes had become the top mathematician in France, whereas Beeckman's mathematical training was poor and his talent limited. Beeckman knew this, just as he knew that not everything in his notebooks was of high value. However, the issue was not the state of Descartes's relationship with Beeckman in 1629 or 1630, but rather their relationship in 1618, at a time when Descartes was not yet a top mathematician or natural philosopher. The real motive for Descartes's anger thus must lie elsewhere.

There are basically two ways to explain Descartes's behavior. The first one is psychological: the independently minded Descartes could not abide the thought that in the past Beeckman had guided him on his route to his new philosophy, even though he was no longer his equal. And so Descartes had to repudiate Beeckman's influence in order to claim complete independence. It is also possible, as Floris Cohen has suggested, that Descartes's accusations of Beeckman's lust for praise are a typical example of psychological projection, because "the obsession with 'praise' and 'being taught' is obviously Descartes's own."[104] If Beeckman once had been a father figure to Descartes, Stephen Gaukroger even suggests that Descartes's attack on Beeckman had a lot to do with Descartes settling scores with his own father, with whom he had had a troubled relationship.[105]

Or perhaps—and this is the second way of explaining Descartes's behavior—Descartes believed that Beeckman posed a threat to his project of fundamentally renewing philosophy. Descartes must have realized, just as Mersenne had, that the *Journal* contained observations testifying to Beeckman's acute insights in physics, which if transformed into a systematic theory of mathematical physics, could change the face of natural philosophy. In 1628–29, Descartes had read in the *Journal* that Beeckman had made an intensive study of the physical astronomy of Johannes Kepler and that he had plans to produce a mechanistic (i.e., a physical-mathematical) interpretation of Kepler's theories.[106] By 1630, Descartes also planned to write a comprehensive account of natural philosophy. In August 1629 he had received a copy of Scheiner's report on the parhelia from Gassendi, which had inspired him to interrupt his work of metaphysics to write a little treatise on meteorological phenomena. He wanted to publish it, as he wrote to Mersenne on October 8, 1629, "as a specimen of my philosophy and to hide behind the canvas in order to listen to what people will say about it" (an allusion to a story about the ancient Greek painter Apelles).[107] Descartes's announcement was made in the same letter to Mersenne in which Descartes expressed anger over Beeckman's presumed misconduct and pretension to be Descartes's master. This cannot be a coincidence. Descartes must have realized that if Beeckman published his ideas, he might appear to be Descartes's competitor or source of inspiration, a circumstance the prickly Descartes wanted to avoid at all costs. Descartes's rude treatment of Beeckman may thus have been a preview, as it were, of a struggle over priority that never broke out openly. Perhaps a gentler scholar than Descartes would not have felt threatened by a friend's plans to publish a similar tract. Descartes, however, had traveled to the Dutch Republic planning to overhaul all existing philosophy, beginning with certain fundamental insights that he alone had acquired, and which, with the help of a method he had developed, would inevitably lead to a strong and indubitable new system. He did not want someone—using alternative methods—to arrive at similar results; nor could he bear the idea of anyone believing that such a colleague had influenced his ideas in any way. That would damage seriously the image that Descartes wanted to create of himself as the philosopher who single-handedly transformed the system of philosophy. Thus, he used a combination of intimidation and contempt to prevent that from happening.[108]

Whether an explanation based on professional rivalry is better than the merely psychological one ultimately depends on how we evaluate the similarities between Beeckman's and Descartes's ideas. If such similarities were real and substantial, then the first explanation is more plausible than the second. This is a discussion that has to be postponed until later. Whatever the explanation, however, Descartes succeeded in disciplining or subduing Beeckman. The two philosophers did indeed even achieve a reconciliation of sorts. After a row, Descartes in the end always calmed down, and Beeckman was not the type to nurse grievances. Thus, Descartes called on Beeckman in Dordrecht in September 1631, while Beeckman had lunch with Descartes in Amsterdam the following month.[109] They also corresponded occasionally.[110] But their relationship was never cordial again, and the trust that normally arises between friends had been irretrievably lost. Descartes's violation of his trust caused Beeckman much grief. Beeckman complained of having few true friends, and that one of them should have abandoned and abused him in this manner struck him as extremely sad. "On the loss of his friends," he wrote in July 1631, "a man can console himself by reflecting that he is now exactly the same as someone who is born without friends. Since that person is not sad because he never had them, why should I be sad for not having them any longer?"[111] The loss of a friend like Descartes may well have exacerbated Beeckman's loss of self-confidence that is perceptible in the mid-1630s.

THE FINAL YEARS

Shortly after his row with Descartes, Beeckman indeed gave up his effort to synthesize his diary and write a treatise on his mechanical philosophy. Furthermore, Beeckman's loss of confidence, or perhaps more the loss of a sense of direction, is demonstrated by his discontinuation of his *Journal*. Beeckman had been in the habit of writing up every idea that he found worthy of note since his student days in Leiden. Sometimes, he made few or no entries for months at a time owing to life circumstances, and sometimes his draft notes were lost.[112] Yet in general he kept up the *Journal* well; there are no substantial breaks. This changed in 1634. Increasingly, his ordinary notes were overshadowed by re-marks on the grinding of lenses, an occupation that became a real passion for him that year. From the end of 1634 onward, he recorded notes only on lens grinding. Beeckman stopped making notes completely, apart from an occa-sional personal remark, after November 9, 1635.

Beeckman does not explain why he stopped writing in his *Journal*. He had not lost interest in natural philosophy: in 1637 he and Colvius were still making astronomical observations from the little tower of the Latin school.[113] Cornelis de Waard's hypothesis that the draft entries after 1634 were lost is improbable, because the character of the *Journal* changed gradually, not abruptly.[114] Most likely, Beeckman's personal circumstances led him to discontinue the *Journal* in these years, as he became more careless and less concerned with external matters. He seems to have become a melancholy man, a sloppy and somewhat sad figure who paid less attention to decorum. "I do not care for my clothes," he noted as early as October 1631, "and I am most negligent in keeping order in my library and my museum."[115] He probably continued to do his duty in the eyes of the outside world, but without conviction, without ambition.

Beeckman must have been a very lively and talkative personality in his youth, but in his student years, when he suffered a bout of depression or perhaps a spiritual crisis, all that changed.[116] He had become quiet, shy, and withdrawn, timid and introverted, obliging and agreeable in his manners of course, but not really at his ease in the company of strangers. He felt most comfortable with his brother Jacob, who was more of a fighter.[117] However, Isaac Beeckman began to break free of his introverted inclination during his years in Rotterdam, perhaps in part because of the proximity of his younger brother. He regained his self-respect and radiated a kind of self-confidence that had been missing for many years. In the conflict with the preachers of Rotterdam, he boldly stood his ground, and he was unquestionably the leader of the Collegium Mechanicum. He was still full of new ideas and initiatives when he moved to Dordrecht. He kept the *Journal* on loose sheets until 1628; then he had it bound in calf, in a clear indication of self-confidence.[118] International scholars' interest in his ideas during his first years in Dordrecht—the happiest time of his life—must also have enhanced his self-respect.

Nevertheless, Beeckman was not spared the cruel blows of fortune. Jacob Beeckman died in 1629, and the break with Descartes came shortly afterward. He endured other losses as well: his wife bore seven children, but five of them died at a very young age.[119] Only in the early summer of 1631, after three infants had died, did it seem that an end had come to his woes. At that moment he had a seven-year-old daughter, Catelyntje, who pleased him greatly; he also had a year-and-a-half-old toddler, Jacob, truly a beautiful child; and his wife was pregnant again.[120] But the longed-for joy was not to be. On July 4, 1631, his little boy died, and the child born in February 1632, a boy named Abraham, lived less than six months.[121] After his son Jacob's death, the bereaved father wrote: "The loss of father, mother, uncles, brothers, and sisters is ordinarily compensated and consequently forgotten through occupations and the pleasure that one has with one's children. Those who after the deaths of their parents do not get any children or lose them, are necessarily the saddest of all."[122] Beeckman, weary of life, considered giving up his position as principal of the Latin school at Dordrecht in 1635, and withdrawing to Zeeland, where most of his family then lived.[123]

The last note in the *Journal* that is not related to lens grinding or matters of personal health is dated November 20, 1634, a dark day for the Latin school in Dordrecht.[124] The plague had struck Holland for a year, and in November the fearful disease penetrated Dordrecht. One of the pupils of the Latin school was among the victims.[125] After it was confirmed that the boy had died of plague, there was a mass exodus from the school. Boys from outside the city who had lodged with the rector received permission to leave to avoid the disease. Seven boys left by ship for Veere in Zeeland, but the ship was run down by another ship halfway between Dordrecht and Middelburg, and six of the seven boys died, "drowning their learning in the water after having rescued it from the fire [the fever caused by the plague]," as a seventeenth-century historian noted.[126] Among the dead were two of Beeckman's nephews.[127] Shortly afterward, one of the teachers in Dordrecht also died of the plague.

With the plague epidemic and the departure of the boarding students— normally around sixty—school life was in serious disarray. The city administration took measures to restore education to its former glory when the crisis abated somewhat toward the end of the year. On January 5, 1635, the council decided to establish an illustrious school, an institution that offered higher education without having the right to grant degrees (a privilege that was limited to universities established by the provincial estates).[128] Deventer had established such a school in 1630, Amsterdam and Utrecht had followed in 1632 and 1634, and the regents of Dordrecht did not want to be left behind.[129] They appointed a professor of history and rhetoric, a professor of Greek and physics, and considered naming a third professor, of mathematics. Naturally, the regents considered Isaac Beeckman for the last position, but he hesitated. In 1633 Isaac had promised his brother Abraham that he could succeed Isaac as rector if he left for any reason.[130] Accepting the professorship would thus help Abraham getting a better job. But Abraham Beeckman had just left for Gorkum in January 1635, and so this particular reason for accepting the offer was no longer relevant. Moreover, the new position itself was not attractive to Isaac, who was just then wishing to be freed of all worries. He finally decided not to accept the offer; the physician Gomarus Walen was then appointed professor of mathematics in January 1636.[131] Hopes that academic life in Dordrecht could be revived by founding an illustrious school quickly faded. The plague appeared to be waning in early 1635, but it returned with greater force later that year and persisted until well into 1637.[132] The epidemic claimed victims from all levels of society, sparing neither the Latin school nor the illustrious school. In 1635, 1636, and early 1637, three teachers died of the plague, along with the three recently appointed professors of the illustrious school. During this period, normal school life was out of the question.[133]

In these circumstances, Beeckman had plenty of time to devote to his new passion, grinding lenses. In 1622 he had been advised by Lansbergen in Middelburg "that I should do my best to make a telescope like the one it appears that Galileus a Galilaeo has had in *Nuntio sidereo* [The Starry Messenger]."[134] But lens grinders in Middelburg and The Hague could not supply the lenses he needed.[135] Over the years, Beeckman became more and more interested in lens grinding. He studied the literature on the subject, consulted a preacher in Bergambacht (whose father had been an alchemist) regarding the materials needed to fix the lenses into the tube, and discussed grinding techniques with a spectacle maker in Dordrecht.[136] At that time, Beeckman did not grind lenses himself. He had not made the telescopes on his little tower: he had borrowed them from others or bought them from an instrument maker in Delft.[137]

The number of *Journal* notes on lens grinding increased significantly in 1632 and 1633, and in 1633 Beeckman bought his own grinding basin. He obtained further knowledge of materials and the technology of lens grinding in Middelburg, which was then still a center of the glass industry. The spectacle maker Hans Lipperhey had invented the telescope in Middelburg back in 1608.[138] By 1633, Lipperhey had died, but Beeckman still could learn the craft from other lens grinders, especially a young man called Johannes Sachariassen.[139] An English lens grinder in Amsterdam also gave Beeckman a few lessons.[140] This teacher was very strict: "He made me sweat," Beeckman wrote.[141] But Beeckman evidently believed learning the craft was worth the trouble. Surprisingly for someone who claimed to have bad eyesight, lens grinding had become an obsession for the erudite principal of the Dordrecht Latin school. Beeckman expressed frustration with his slow progress in learning the craft. In 1634 he noted that he had not got very far: "[I] go to it much too sloppily: stuff hangs on the clothes, on my hands, arms etc., which I lay down here and there as I am washing glass, not shaking it off well enough, not washing my hands, touching my dusty clothes with my arms."[142] By the end of 1634, he finally began to master the technique and, by the spring of 1635, even dared to enter a competition with Sachariassen, which he won, thanks to an improvement in grinding methods that he invented. When they compared their lenses in a dark room at the home of Jacob van Lansbergen, Philip's son, the lens made by Beeckman turned out to be the best.[143] Yet, as far as we know, he never made a telescope. Perhaps Beeckman was not only working on making a telescope. He may also have been working on a microscope, hoping to actually see the small particles of matter that he had postulated in his corpuscular philosophy of nature.[144] As mentioned before, in March 1631 he had copied a letter written by Cornelis Drebbel in 1613 to King James I of England and added the drawings of several of Drebbel's instruments, including a compound microscope. It may have inspired him to pursue the grinding of lenses himself.[145]

Beeckman visited friends and acquaintances, as well as the English spectacle maker, in Amsterdam. Justinus van Assche, Beeckman's brother-in-law, lived on the Prinsengracht, and they maintained a warm relationship, despite the fact that Van Assche was a Remonstrant.[146] Descartes lived on the Westermarkt, and Beeckman visited him at times, too. Beeckman also visited the map maker and printer Willem Jansz. Blaeu, who had a shop near the harbor which was an obvious meeting place for those interested in mathematics and related subjects, including astronomy, cartography, and the art of navigation.[147] Martinus Hortensius, the only one of Beeckman's pupils to achieve anything in the mathematical sciences, also lived in Amsterdam.[148] Hortensius had just been appointed a professor at the Athenaeum illustre of Amsterdam when Beeckman visited him on August 1, 1634. He was a promising astronomer; besides having studied with Beeckman, he had worked with the Leiden astronomer Willebrord Snellius and for a time served as Philip van Lansbergen's right-hand man. Hortensius also maintained an extensive network of correspondents as far as Italy and southern France. Upon receiving permission to teach courses in mathematics in Amsterdam on March 9, 1634, Hortensius appeared on the verge of a brilliant career. Beeckman could be justifiably proud of his former pupil. Hortensius gave Beeckman one of the two copies of Galileo's *Dialogo* he had acquired; in return, during his visit, Beeckman allowed Hortensius to inspect his *Journal*. Hortensius was only the third person, after Mersenne and Descartes, to attain the privilege of studying Beeckman's notebooks.[149]

Partly through Hortensius, Beeckman became involved in his final years with the discovery of a reliable method of determining geographic longitude at sea, which was of the utmost importance to a seafaring nation. This problem persisted since the Europeans had begun their exploratory voyages.[150] To determine latitude, it sufficed to observe the position of the sun at midday, or the Pole Star at night; but there were no such simple methods for measuring longitude. Some believed that longitude could be determined from the magnetic needle's declination from true north and one or more tables. Others believed that the difference between local time (as measured by the position of the sun) and the time in the home harbor could be marked by a good clock. In principle, this solution was correct, but at that time there were no clocks that could keep time accurately and resist the motions of the ship. Lacking an accurate mechanical clock, sailors might have used a celestial clock to solve the problem. If they knew when and at what time a particular celestial event would take place, or when a certain constellation would appear, they could establish an absolute time and compare it with the local time aboard ship. Galileo believed he had found a celestial clock: in his opinion, the rapidly changing positions of the moons of Jupiter, which he had discovered in 1610, offered a possible solution. He resolved to make a table of the positions of these moons and to give copies to navigators. However, the king of Spain, to whom Galileo offered his discovery in 1612, was not impressed and negotiations came to naught. Thereafter, Galileo devoted himself to other matters.[151]

Many years later, however, Galileo became aware that the States General of the Dutch Republic had offered a large sum of money to anyone who found a practicable solution to the problem of determining longitude. At first, he noted the offer and then ignored it, but in 1633 when he was censured for his outspoken Copernicanism and forced to pursue less controversial subjects, the further development of his solution to the longitude problem presented itself as an opportunity. At the same time, Galileo's friends considered possible ways of extracting the condemned natural philosopher from Italy and providing him with a refuge in the Dutch Republic. Hugo Grotius—an established scholar, diplomat, and a former leader of the Remonstrants in Holland who then lived in exile in Paris—was able to draw the interest of prominent Amsterdam citizens, including Hortensius, Blaeu, and Laurens Reael, a former governor-general of the Dutch East India Company, in such a project.[152] Perhaps, they considered

Galileo for a position as professor at the Athenaeum illustre. Grotius reported in July 1635 that the Italian knew a good method for determining longitude at sea, as an added argument for bringing Galileo to the Netherlands.[153] But a month later, Grotius informed his Amsterdam correspondents that Galileo could not leave Italy on account of his advanced age. Yet, by enlisting the aid of Elia Diodati, who was an advocate in the Parlement of Paris and a good friend of Galileo, Grotius tried to persuade the Italian master to present his method officially to the States General.[154] Hortensius already had begun correspondence on this matter with Diodati and Galileo.

On August 15, 1636, Galileo indeed offered his method of determining longitude to the States General.[155] Galileo's friends in Amsterdam and The Hague were already informally aware of this offer. On November 11, 1636, when the States General received Galileo's letter through Reael, who was regarded as the major advocate of Galileo in Holland, they immediately appointed a committee to look into the matter.[156] The members of the commission were Reael, Blaeu, and Hortensius. They received permission to add Jacob Golius, professor of mathematics at Leiden, as a possible fourth member.[157] In place of Golius, however, Beeckman was added to the commission, apparently on the recommendation of Hortensius, who knew that in 1631 Beeckman also had proposed utilizing the moons of Jupiter as a celestial clock, independently of Galileo.[158]

Reael was chair of the committee, but Hortensius—who wrote letters to Diodati, Galileo, and others involved—was its most active member. Yet Diodati found that the whole business was taking too long; moreover, he was upset that the matter, so carefully prepared, had been leaked, through Beeckman and Mersenne, to the French astronomer Jean-Baptiste Morin, one of Galileo's competitors.[159] The Dutch tried to reassure Diodati. Constantijn Huygens, secretary to stadholder Frederick Henry, explained to him that the committee had to judge not only the theoretical correctness of the method but also its practical application and that would take time. For example, they had to make tables before they could tell if the method was accurate enough.[160] Meanwhile, as a token of their goodwill, the States General gave Galileo a gold chain valued at five hundred guilders.[161] In a lengthy letter dated April 27, 1637, Hortensius assured Diodati that no essential details had been passed on to Morin; thus there was no reason to fear his competition.[162]

Still, there were other reasons for concern. The letter in which Reael officially told Galileo of the provisional reaction of the States General did not reach the Italian scientist until June 23, 1637.[163] When Galileo wrote a further detailed letter on August 22, giving instructions on how to use instruments in the investigation, it was too late.[164] The letter reached Reael, but then remained unopened on his desk. On October 25, the councillor Nicolaes van Reigersberch informed his brother-in-law Hugo Grotius that Reael had died two weeks earlier, on October 10. "The loss of two children through the contagious sickness [the plague] had plunged this good man into such profound melancholy, that he forgot all other thoughts and even those that were very dear to his heart. Even a letter from Galileo Galilei, which was given to him when he was still healthy, remained unopened, which I mention so that the said Galileo may be informed of it."[165] At this point, the matter reached a dead end. Blaeu died in October 1638; and Hortensius, who was planning to visit Galileo in person, died a year later, on August 17, 1639. The States General viewed the matter as closed. They briefly considered convening a new committee in 1640, but nothing came of it.[166]

BEECKMAN'S DEATH

And what happened to Beeckman? Beeckman had been the first committee member to die in May 1637 when the discussions about Galileo's offer were still underway. His death was not unexpected. In his immediate surroundings, he had seen various people die of tuberculosis: his brothers Gerson and Jacob, his brother-in-law Abraham du Bois, and his Rotterdam friend Gerard van Berckel. Isaac had allowed an autopsy after the death of his brother Jacob, and he had closely monitored the illnesses of Du Bois and Van Berckel.[167] He was continually preparing for his own death, and more than anything else was afraid of being buried alive. He knew the symptoms of tuberculosis too well, and when he suffered a rapid and striking weight loss in the first months of 1637, he understood what he was up against.

The first sign of consumption occurred on May 15, 1636, when Beeckman spit up blood.[168] Yet he did not stop his busy round of activities. He traveled to Middelburg to take care of family business, became a member of the commission to investigate Galileo's offer, and continued to make astronomical observations until March 1637.[169] On May 9, however, Beeckman noted that from January to May his weight had declined from 118 to 106 pounds, whereas his normal weight was 125.[170] On May 19, 1637, Beeckman died at the age of forty-eight, leaving a wife and two underage children.[171]

We know of few reactions to the news of Beeckman's death. When Colvius informed Descartes, who had more or less reconciled with Beeckman, the French philosopher sent a rather cool letter of condolence. On June 14, he wrote he was certain that Colvius would be distressed. "However, Sir, you know much better than I do that the time we live in this world is so short in comparison with eternity, that we should not worry ourselves too much if we are taken a few years earlier or later. Since Mr. Beeckman was extremely philosophical [extremement philosophe], I have no doubt that he had been resigned for a long time to what has happened to him. I hope that God enlightened him so that he died in His grace."[172] In Descartes's *Discourse on the Method*, published less than a month after Beeckman's death, Beeckman is not mentioned at all. Thus, Descartes effectively obliterated the formative influence of Beeckman, who first inspired him with his truly novel mathematical-physical philosophy of nature. John Cole, Descartes's biographer, rightfully calls it "the most serious shortcoming of the *Discourse* as an intellectual autobiography."[173] Mersenne did not learn of Beeckman's death until May 1638, more or less by chance.[174] Whether Diodati indeed told Gassendi, as Hortensius had asked him to do in a letter of June 22, 1637, is unknown.[175] In Dordrecht, a new principal replaced the deceased rector in March 1638, and around that time his widow and her two daughters, Catelyntje and Susanna, moved to Middelburg.[176] The younger child, Susanna, died there in October of that year.[177] Only Catelyne reached adulthood and later married a regent of the Zeeland harbor city of Vlissingen.[178]

On July 14, 1637, two months after Beeckman's death, his library was sold at a public sale.[179] Usually, the widow of the principal was entitled to a "year of grace," during which she was paid her husband's salary and was allowed to stay in the house where they had lived. She had the opportunity to arrange her affairs during this period, and this probably entailed the sale of Beeckman's extensive library. The catalog, as printed by the Dordrecht bookseller Isaac Andreae, is a remarkable testament to Beeckman's wide-ranging interests, especially when compared to the books mentioned in the *Journal*. Of course, the contents of an auction catalog do not exactly reflect a person's library. The owner may have given away certain books shortly before he died or members of his family may have decided for one reason or another to leave certain books out of the sale. This must have been the case with Beeckman's copy of Galileo's *Dialogo*, which he valued highly, for it is not in the catalog.[180] Furthermore, booksellers were in the habit of adding other books they had for sale, and they also left out items they believed would not sell.[181] Still, with all these caveats taken into account, the catalog is an interesting document. The relative scarcity of mathematics and natural philosophy books is revealing. The books listed (566, duplicates included) fall into the following categories: classical and neo-Latin literature (with many editions of Terentius), 36 percent; theology, 28 percent; medicine, nearly 9 percent; history, both ancient and contemporary, nearly 7 percent; logic, 4 percent; general philosophy, 4 percent; and natural philosophy and mathematics, each comprising only 3.4 percent.[182] The theology section is rather ecumenical: Calvin is represented with at least eleven titles, but we also find books written by Thomas Aquinas, Bellarmine, Erasmus, Luther, Melanchthon, Gisbertus Voetius, Teellinck, and even H.N., which stands for Hendrik Niclaes, the leader of the spiritualist sect of the Family of Love. The list is more selective in mathematics and natural philosophy: Beeckman owned the complete works of Aristotle both in Greek and in Latin; he had three or four editions of Lucretius, three titles of Bacon, and two of Fludd, but the complete works of Galen are missing. Euclid is represented only in the Dutch translation of 1606, only Stevin's *Art of Fortification* is on the list, and whereas Cardano's *De subtilitate* is listed, J. C. Scaliger's famous commentary is not.[183] Of course, the medical section of the catalog contains some references that are also of interest for natural philosophy, but the works of Copernicus, Kepler, and Galileo are missing. The Beeckman we meet as we go over the catalog is first and foremost a schoolteacher, with a pronounced interest in theology and medicine, but with no particular interest in natural philosophy or the mathematical sciences. No one would have guessed the actual role Beeckman played in the early years of the scientific revolution if we had known nothing of Beeckman but this auction catalog.

But what happened to Beeckman's *Journal*? Upon learning of Beeckman's death in 1638, Mersenne asked both Rivet and Van Beverwijck about the manuscript, fearing that it would be sold to an apothecary, who would turn its pages in little paper bags in which to sell his drugs.[184] However, Beeckman's younger brother Abraham—who after a short stay at Gorkum had become rector of the Latin school at Vlissingen in 1636—took good care of the manuscript and in 1644 published a selection of his brother's notes as *D. Isaaci Beeckmanni Medici et Rectoris apud Dordracenos Mathematico-physicarum Meditationum, Quaestionum, Solutionum Centuria*.[185] In the preface, Abraham Beeckman explains that he decided to publish excerpts from the literary remains of his brother to defend him against detractors who charged him with wasting time on idle speculations. Abraham had wanted to publish Beeckman's letters, but the manuscript he submitted to the printer disappeared, and the projected book did not materialize. Instead, he presented one hundred notes from Beeckman's *Journal* (both in Latin and in Dutch), promising to publish another hundred in case the first selection would meet with a favorable reception. Abraham Beeckman took the notes from the *Journal* up to 1629, more or less as he found them, grouping them according to their subject, but with very little additional editing.[186] As such, the publication of the *Centuria* was a fine and loving gesture, but it came too late to have any influence on the development of natural philosophy.

Abraham Beeckman added several biographical notes concerning his brother, his sisters, and his cousins to the *Journal*. He briefly described Isaac Beeckman's life and gave a brief impression of his character. We have to make do with Abraham's description in the absence of a portrait of Beeckman. "[He] was short of posture, as his father had been; great in judgment, excellent in understanding, sweet by nature, and agreeable in conversation. He avoided all quarrels and discord; was greatly beloved by his pupils and kind to everyone." And he was, Abraham Beeckman rightly recalled, "always busy speculating, as this book can testify."[187]

CHAPTER FOUR

Principles of Mechanical Philosophy I: Matter

Isaac Beeckman's ideas about the natural world come to us in a chaotic, disordered manner, even in the printed pages of the *Centuria* and the *Journal*. However, these ideas themselves were not chaotic and confused. Beeckman's notes and speculations present a remarkably coherent though not systematically developed philosophy of nature. Although Beeckman's philosophy appears fragmented at times, he had clear ideas regarding the structure of the natural world and the best way to examine and learn more about nature. Reaching this conclusion is not just a matter of historical reconstruction. Beeckman offers repeated summaries of his main ideas, his "principles," as he called them, in his writings. Specifically, Beeckman presents his thoughts regarding the study of the natural world when preparing his 1618 medical dissertation and the oration that was part of the ceremony. Furthermore, he provided Gassendi and Mersenne with an overview of essential aspects of his philosophy during his conversations with them in 1629 and 1630.[1] Of course, Beeckman sometimes changed his mind about particular explanations of natural phenomena, but he never wavered in his ideas about the basic principles of natural philosophy, which he had developed largely in the period from 1612 to 1618. Therefore, Beeckman's philosophy of nature forms a coherent whole, despite the disorder of the *Journal*.

In general, one should distinguish between Beeckman's concepts regarding methodology and his ideas on the constitution of the natural world. Two basic notions emerge regarding the latter: matter and motion. Essentially, Beeckman's philosophy of nature involves a combination of a new theory of matter, grounded in atomism, and a new science of motion, based on the idea of inertia. For the sake of clarity, it is best first to examine his ideas on method and his theory of matter and then to review Beeckman's new science of motion.

METHODOLOGICAL PRINCIPLES

Beeckman sometimes referred to his practices as "speculations." When the burgomasters of Dordrecht agreed to erect an observatory on the roof of the Latin school in 1628, Beeckman noted that the tower was "for my speculations."[2] Abraham Beeckman used the same terminology when he characterized his brother as "always busy speculating."[3] Abraham's characterization was accurate: Isaac Beeckman neither learned about the laws of nature by carrying out purposeful experiments nor conducted systematic "research" in a specific field of natural philosophy. Instead, he devised possible explanations for diverse phenomena by continuously reflecting on how the phenomena of nature *might* be explained.[4]

But a number of fundamental ideas, his "principia" or "ὑποtheses," guided Beeckman in his speculations.[5] These included both methodological principles and some basic notions concerning the constitution of nature. Two main principles dominate Beeckman's methodology. First, Beeckman was convinced that a combination of mathematics and physics constituted the best way to analyze natural phenomena, which leads to his "philosophia mathematico-physica."[6] Beeckman's second principle established that only explanations that the human mind can represent clearly and visually should be allowed in natural philosophy.

Beeckman repeatedly argued for the application of mathematical reasoning to natural philosophy. Furthermore, he declared that mathematics was indispensable to the practice of natural philosophy in his 1618 dissertation lecture.[7] Indeed, Beeckman based his critique of the French physician and natural philosopher Sébastien Basson, who was noted for his opposition to Aristotle, on Basson's lack of mathematical reasoning. In 1623 Beeckman read Basson's *Philosophiae naturalis adversus Aristotelem Libri XII*, published in 1621, and agreed with the atomistic perspective Basson advanced. But without mathematical reasoning, Basson undermined the theory of motion advanced in his book.[8] In 1620 Beeckman criticized some editions of Galen's works for omitting the mathematical figures: "Medicina requiret mathesim," he later added to this note.[9] By contrast, Beeckman found a like-minded spirit in Descartes. Indeed, shortly after their first meeting in 1618 the young French philosopher told

Beeckman that he had never met anyone like him who strove to combine mathematics and natural philosophy.[10] The dynamic derivation of the law of falling bodies that they developed together serves as an excellent example of using mathematics to solve problems in physics. And Beeckman demonstrated his ability to apply physical-mathematical philosophy independently when he explained to Descartes the proof of the inverse relationship between the length of a string and the pitch it produces, which he had developed in 1614.[11]

Beeckman's 1627 inaugural lecture at Dordrecht demonstrates clearly what his physical-mathematical philosophy amounts to.[12] In this lecture, Beeckman shows that many natural phenomena can be explained in a simple manner if one takes into account the geometric properties of the bodies involved. He used what were called isoperimetric figures, closed curves with the same perimeter but different forms and thus enclosing different areas, to illustrate this point. For example, round cities are easier to defend than square or rectangular ones because, given the same length of the walls, they enclose a greater internal area and can thus hold more houses, more people, and more defenders. Circles and spheres are the most efficient: with a minimum of circumference, they enclose a maximum area or volume. Beeckman added that God, the very wise Architect, created the world in spherical form for the same reason.[13]

In the second part of the lecture, Beeckman focused on figures that have the same geometric form but differ in size. For example, when three-dimensional figures are enlarged, their volume increases faster than their circumference or surface area, and the larger of two similarly shaped bodies has relatively more volume than the smaller. Therefore, large bodies remain in motion longer than smaller ones of the same shape in a resistant medium, because deceleration occurs as a result of the object's collision with the medium and is thus directly proportional to its surface area. On the other hand, Beeckman argued, the "moving force" (vis) depends on the "quantity of matter" (corporeitas) and is thus directly proportional to volume.[14] Because the relationship between surface area and volume for larger objects is relatively favorable, such objects can more easily overcome the medium's resistance. Thus, Beeckman explained why the Colossus of Rhodes, a giant statue that in ancient times stood in the harbor of the Greek island of Rhodos, was less likely to be affected by a powerful storm than a human body by a strong wind. He had already noted these examples and others like these in his *Journal*, but he tied them together to illustrate his physical-mathematical philosophy.[15]

Beeckman was not advancing what has been called the mathematization of nature. He did not view mathematics as a language in which reality would have to be described in order to arrive at an exact formulation of the fixed relationships between observed phenomena. Mathematization, as Descartes argued at about the same time in his *Regulae*, is possible only if a particular field of knowledge satisfies certain conditions. The suitability for mathematical treatment arises only when that field is subject to *ordo* or *mensura*, that is, when the statements made about it can be arranged in deductive chains (axiomatization), or when quantitative relations can be established between the observed variables that permit an algebraic treatment (algorithmatization).[16] Neither is applicable to Beeckman's "philosophia mathematico-physica." He accepted the natural world as it presented itself to his senses, concentrated his energies on observed geometric forms of natural bodies, and did not establish mathematical relationships between the separate phenomena. Beeckman's physical-mathematical philosophy served to explain natural phenomena by integrating natural bodies' simple geometric properties with their mechanical properties (primarily movement and inertia).

The physical element clearly dominated Beeckman's physical-mathematical philosophy. Beeckman viewed himself as more of a natural philosopher than a mathematician. Thus, in the derivation of the law of falling bodies in 1618, it is no coincidence that Beeckman took a more physical approach, while Descartes was more mathematical.[17] Furthermore, it is striking that Beeckman, when reading Kepler's works, focused almost exclusively on the physical aspects of astronomy like the causes of the attraction and the mechanisms that lie behind the phenomena, while barely mentioning the mathematical aspects, such as the precise form of trajectories of the heavenly bodies. Finally, it was primarily in his contacts with others that Beeckman was concerned with standard problems in the mathematical science of motion, that is to say mechanics. In conversations and correspondence with Descartes and Mersenne, mechanics at least played a larger role than in the *Journal* itself. Left to his own devices, Beeckman preferred to occupy his time with physics, not with mathematics.

In Beeckman's increasingly complicated corpuscular mechanisms explaining natural phenomena, experience also seems to be of limited importance. Beeckman in no way downplayed the importance of making observations. His lengthy series of systematic meteorological observations in Zierikzee and the establishment of his observatory in Dordrecht testify to his belief in the value of observations. His basic explanatory principles also were, of course, grounded in sense experience—as can be seen when Beeckman used atomism to explain the behavior of water and air. But Beeckman was tenacious in defending principles he had developed, even in the light of contrary experience. In 1630, when

Mersenne presented Beeckman with experiments that appeared to conflict with his concepts in musical theory and acoustics, Beeckman answered that he would prefer to leave such observations unexplained than relinquish principles based on his own experience.[18] Although Beeckman was often timid in his interactions with others, he stubbornly clung to his principles when he was confronted with others' objections to them.[19] Beeckman's remark in a letter to Mersenne that he found it superfluous to establish by experimental means whether a wooden object falls more slowly through the air than an iron object of the same shape and size testifies to Beeckman's limited belief in experimentation. After all, Beeckman asserted, reasoning tells us again and again that it must be so.[20] He reasoned that the resistance of the displaced air increases continuously, so that at a given moment the acceleration of the falling object stops and a certain final velocity is reached. Thus, the heavy iron object—its greater mass can overcome the increasing air resistance better than the wooden object—will reach a higher final velocity and hit the ground earlier. According to Beeckman, experimentation is necessary only to determine the precise rates at which the iron and wooden objects fall.

In the light of the uncertain criteria for trustworthy experiments in the early seventeenth century, Beeckman's seemingly rationalistic position appears quite sensible, but there also was a downside. Controlled observation and serious experimentation might have served to moderate an inventive natural philosopher's theories, which tended to explain too much rather than too little. Beeckman's theory of matter, for example, was so flexible that it could explain nearly all phenomena. Perhaps Gassendi's 1633 praise of Beeckman's "solertia," his ingenuity in thinking of new corpuscular mechanisms that lay behind phenomena, was double-edged.[21] Even Beeckman at times acknowledged that he should pay more attention to experience—for example, in 1620 when he considered the vexing problem of atoms' capacity to rebound, which conflicted with his principle that atoms were perfectly "hard" (solidus) and inflexible.[22] Beeckman momentarily agreed with Galen, who had said that (in natural philosophy) one should not stick too much to one's principles because they are always subject to philosophical dispute; one should concentrate instead on phenomena that can be observed by the senses. Nevertheless, Beeckman, for didactic and pedagogical reasons, could not abandon his principles easily. "He who is able to arrange a certain argument from the first principles to its ultimate components," he wrote, "completes his full course a lot quicker [compendiosius], putting everything in its right place, using an efficient method for all things, not much different from those who store the same things in their memory with the help of pictures and imaginary places."[23] His principles therefore served as more of a mnemotechnical device, so popular in the sixteenth and early seventeenth centuries.[24]

Despite Beeckman's attachment to his fundamental principles, his main criterion for assessing the soundness of a particular explanation, clearly stresses the role of the senses in acquiring knowledge. Thus, only those explanations that allowed the human mind to form a mental picture of the mechanism that was behind the phenomena—literally, to "imagine" what was going on—were acceptable according to this basic rule of philosophical reasoning. When defending his dissertation in 1618, Beeckman formulated this precept as follows: "In philosophy we must not allow a single regulation, or a single prescription or a single rule which has not been approved as absolutely true and completely certain by reason and which does not appear to the mind as open and 'naked' (nude) as visible things do to our eyes."[25] The element of "picturability" (the German word *Anschaulichkeit* and the Dutch equivalent *aanschouwelijkheid* express this concept more concisely but are difficult to translate) plays a crucial role. Beeckman emphasizes this concept in an October 1629 letter to Mersenne: "In philosophy I allow nothing that is not represented to the imagination as if it were observable" (Nihil enim in philosophia admitto quam quod imaginationi velut sensile representatur).[26] For Beeckman, the restriction of natural philosophy to what the mind can grasp through the senses, particularly through the eye, derived from the idea that God had given man the power to penetrate and learn the true nature of the world. He referred to his notes about the question of why bees build their combs in hexagonal forms, "attributing to God the construction of all nature, which He had done in such a way that he subjected to our understanding the complete contemplation of all inferior things that have been made by nature."[27] When the true causes of a phenomenon cannot be determined immediately, Beeckman argued that we should maximize our use of analogies to what is known and understood.

For Beeckman, a direct consequence of this axiom was that the philosopher must make a clear distinction between spirit and matter, and must be careful not to explain material phenomena by spiritual and thus preeminently nondemonstrable causes. Anyone who violated this precept would get entangled in occult explanations, resulting in confusion. Beeckman cites the attribution of intelligence to material objects such as planets, or the acceptance of the idea of a world soul as examples of the errors produced by violating this axiom.[28] He resolutely rejected Kepler's notion of a "formative power" in nature out of hand: "A formative power in nature, or better, in heat, strikes me as very foolish and unworthy of a philosopher. That is not offering a cause, but hiding one."[29]

Beeckman also believed it was impossible to speculate about the influence of spirit on matter. How could something immaterial cause something material to move?[30] How should one imagine that happening? Beeckman allowed for one exception, namely the way in which people use their will to influence the movement of their bodies, but he found it problematic to incorporate this phenomenon into his system: "Is it not strange, that a man can move his hands, feet, and limbs, as he wishes, through his thoughts?"[31] For Beeckman, this constituted the only exception to the general rule that the material cannot be influenced by the immaterial.

The corollary to this principle is that one object can act upon another only by means of physical contact. Forces that work at a distance, such as magnetism, cannot be allowed in natural philosophy as such, "because all force works through contact."[32] Nothing in nature occurs "absque corporali contactu" or, in Dutch, "sonder rakinge" (without touching).[33] Forces that appear to work at a distance therefore must be viewed as the result of material mechanisms invisible to the naked eye.

Thus, Beeckman viewed the world as consisting of a large number of small material particles that move in empty space and are distinguishable from each other only by their form, size, and state of motion. According to a note from 1618, the physical world comprised matter in motion.[34] "All forces," Beeckman stated, "arise from movement, form, and quantity; and therefore these three things must be considered in every problem."[35] Only matter and motion resist reduction to something else: "Who would be able to understand what it is that causes resistance, and whence the first motion came?"[36] On the other hand, colors as well as forces, heat, and other secondary qualities are only the visible effects of corpuscles in motion and not true characteristics of matter.[37]

Given these outlines, Beeckman's philosophy of nature may be safely called mechanical. One of the definitions given by E. J. Dijksterhuis, still one of the great authorities as far as the history of the mechanical philosophy is concerned, fits Beeckman's fundamental ideas very well: it is the philosophy that "assumed no other explanatory principles besides matter and motion and no other way in which material bodies influenced each other besides contact."[38] For sure, Beeckman did not refer to his philosophy as a mechanical philosophy of nature. Henry More and Robert Boyle introduced this notion in the late 1650s only to characterize Descartes's corpuscular philosophy of nature. According to Boyle, a mechanical philosophy is the notion that "the Universe being once fram'd by God, and the Laws of Motion being settled and all upheld by His incessant concourse and general Providence, the Phaenomena of the World thus constituted are Physically produc'd by the Mechanical affections of the parts of Matter, and that they operate upon one another according to Mechanical Laws."[39] This rather late arrival of the term *mechanical philosophy*, however, has not kept historians of science from using the term in the case of Descartes, and there is no reason not to use it in the case of Beeckman.

The fact that the term *mechanical philosophy* was coined nearly half a century after Beeckman had articulated some of its basic precepts is no coincidence. In the first half of the seventeenth century, the term *mechanical* in common parlance still referred to the mechanical arts and to manual operations (including chemical operations) that were deemed unworthy of a philosopher. Alan Gabbey, who studied the original meanings of the word *mechanical* in several languages, found that the most common meaning in French was "mean, cheap, miserly."[40] It could also mean geometric (when used in the context of certain instruments) or simply technical (when used in the context of simple machines). In 1618, when Beeckman mentioned his own "negotia mechanica," he specifically referred to his hydraulic engineering.[41] Yet the association with "getting one's hands dirty" was never far off.

Therefore, it is not surprising that the only time Descartes referred to his "rather plump and mechanical philosophy" (pinguiuscula et mechanica philosophia), he used the term ironically.[42] In 1637, commenting on Descartes's *Discourse on the Method*, Libert Froidmont—a professor of theology and philosophy at Leuven University—had accused Descartes of copying the physics of Epicurus, which he called "coarse and rather plump" (ruda et pinguiuscula). In particular, he had criticized Descartes's *Météores* for its account of the composition of earth, air, water, and other bodies, which in Froidmont's eyes was "excessively gross and mechanical" (nimis crassa et mechanica). Descartes pretended not to understand the basis of Froidmont's complaint when he wrote to another professor from Leuven, Fortunatus Plemp: "If my philosophy seems to him excessively gross because it considers shapes, sizes, and motions, as happens in mechanics, he is condemning what I think deserves praise above all else, and in which I take particular pride."[43] To Descartes, mechanics—despised by the old-fashioned philosophers, but welcomed by the mathematicians—is a part of physics and thus a part of philosophy, and as such much truer than the other parts of philosophy, because it relates, as Descartes adds, to use and practice. "So that if he is belittling my way of philosophizing because it resembles mechanics, that seems to me the same as if he were to belittle it because it is true."[44]

Still, Descartes never again referred to his philosophy as "mechanical." Apparently, the association with mean and cheap lingered. Froidmont's criticism would certainly not have impressed Beeckman, who was not afraid of getting his hands dirty. Yet Beeckman might not have used the term either, for marketing reasons. He preferred to call his view of nature a "philosophia physico-mathematica" (or "mathematico-physica"). Beeckman's philosophy of nature, however, meets all the criteria for identifying it as a mechanical philosophy. Further examination of the finer points of his thoughts will demonstrate the truly radical dimensions of Beeckman's mechanical view of nature.

THE STRUCTURE OF MATTER

How did Beeckman specifically think about the structure of matter? Was he really a consistent atomist, and did he believe in the existence of indivisible particles of matter in a vacuum? Or did he refuse to commit himself to any specific theory, limiting himself to accepting corpuscles, divisible or not? Were these bits of matter all the same or did they differ in form and size? What sort of new explanatory possibilities did they open up?

Letters Beeckman wrote in 1613 to Jeremias van Laren—a fellow Zeelander studying in Franeker at the time—contain his first coherent statements on the structure of matter.[45] One question that emerged in their correspondence was whether vacuums exist. Beeckman defended the view that the presupposition of a vacuum was necessary in order to explain the existence of movement in the world.[46] He presented Van Laren with the following syllogism: "In a continuum or a world in which all things are on all sides contiguous, nothing moves. Yet in the air there are birds, and in water fishes do move; therefore the things in the world are not contiguous on all sides."[47] If the world were a continuum, there would not be room for movement. Imagine a row of "little balls" (globuli): the ball in the extreme left position could move to the right only if the ball on the extreme right could move farther to the right in free space. There would be no free space in a world entirely filled with substance, and thus starting from a resting position, no motion could occur. Also, if the world were a continuum, movement in one part would be instantaneously transmitted to the other end of the world, which is plainly absurd. Finally, in case one would argue that air is contiguous and still can be compressed, this too can be explained only if one posits the existence of a vacuum, empty space *inside* the particles of air.

Van Laren took a different view. God had given matter the "capacity to become more or less dense" (potentia rarescendi et densandi), and therefore air could occupy differing volumes without necessitating a vacuum.[48] Beeckman disagreed, believing "that humans should not invoke the will of God in such cases because they know nothing about His will" (cujusque modus inexplicabilis est menti humanae). In this way, one might give diametrically opposed explanations of the same phenomenon.[49] As Beeckman noted in his *Journal*, the fact that bodies move presupposes the existence of the vacuum.[50]

While Van Laren argued for the Aristotelian tradition, Beeckman endorsed a contrasting classical tradition, which can be traced back to Hero of Alexandria, the famous engineer of the first century A.D.[51] In the introduction to his *Pneumatica*, a book about pneumatic and mechanical devices, Hero had given a brief explanation of the structure of matter as he understood it on the basis of his experiments with the apparatus described. His presentation did not go further than a simple particle theory, which certainly does not deserve to be called atomism. Hero attributed to air and water particles only those characteristics that he needed for an explanation of such varied phenomena as the compression of air, the sucking up of water in a pump, and the expansion of vapors. He conceived of matter as a system of hard physical particles between which remained empty space, the "vacua intermixta."[52]

Beeckman had become familiar with these ideas during his student days in Leiden. As a candlemaker's son and someone who installed water supply systems, he undoubtedly found them useful. Hero's name figures on the list of authors that Rudolph Snellius had recommended to him in 1608 in order to broaden his knowledge of mathematics.[53] Yet he probably did not read Hero right away, because in his subsequent notes he referred only to Cardano's extremely popular *De subtilitate*.[54] While discussing a fountain, Cardano observes that Hero had introduced the "vacua intermixta" as the explanation for the working of its mechanism.[55] Beeckman did not appear to have a copy of Hero's work at his disposal until January or February 1619.[56]

By that time, however, Beeckman's still rather unspecific particle theory had evolved into a full-fledged and quite radical atomism. In late 1614, Beeckman mentions the famous Latin atomist Lucretius, whose explanation of magnetism appealed to him.[57] But more than anything, Beeckman's study of medicine (mainly from 1616 to 1618) appears to have contributed to the development of his atomistic theory of matter. Beeckman learned about the theories of several classical atomists through Galen's works (which he did not own, but must have borrowed from someone else, probably Lansbergen). Galen disagreed with these atomists, but Galen nevertheless presented their theories clearly. Beeckman greatly admired Galen's medical knowledge, but he regretted the vehemence with which the ancient physician had argued against the Greek atomists. As he saw it, Galen's denunciations of Democritus, Epicurus, and Asclepiades only testified to his deep ignorance of their opinions.[58] However, Galen's writings sparked Beeckman's curiosity about their ideas. Beeckman's interest in pharmaceutics, already discernible during his student years, also must have stimulated his corpuscular speculations. He definitely became an atomist in the course of his medical studies. Beeckman declared that the vacuum intermixtum does indeed exist in one of the corollaries attached to his thesis. Furthermore, in a contemporary *Journal* note—a reply to an anticipated objection—Beeckman stated that all things consist of atoms that have different shapes and that are found at varying distances from each other, with empty space in between.[59] Differences in form and spatial arrangement of atoms make some materials opaque, while others can be polished so smoothly that they reflect light, and still others are completely transparent.

As Beeckman had speculated in 1617, there were only four kinds of atoms, which could be distinguished from each other only by their geometric form and were all made of the same inert material.[60] These four atoms are analogous to the four Aristotelian qualities of heat, cold, moistness, and dryness, corresponding to the four elements earth, water, air, and fire. Of the senses, it is touch that can best apprehend these qualities, and touch is also most sensitive to *figura* or (geometric) form. Now, the basic forms of atoms could vary within a certain range (e.g., some atoms of a particular kind might be a little larger than others of the same kind), but there were no intermediate forms.[61] One could not be absolutely certain that there are not more than four kinds of atoms; Beeckman knew that some chemists worked with only three basic elements. But, in any case, no more than four different forms were necessary. For example, light and the sun can be conceived simply as comprising fire atoms; thus there is no need to consider light as a fifth kind of atom, which would correspond to the fifth element, ether.[62]

Beeckman's atomism at this stage is based on the Aristotelian system of elements and is in fact an atomistic rewriting of that system. The fundamental importance that Beeckman assigns to the sense of touch—primarily conceived as the capacity to experience resistance and to distinguish forms—echoes the manner in which Aristotle had based his theory of matter on tactile observations. And Beeckman's coupling of the four atoms with the Aristotelian elements is likewise striking. Such reliance on Aristotelianism was not uncommon in the sixteenth and seventeenth centuries: various interpretations of the peripatetic system of the elements look like embryonic particle theories. The system of "minima naturalia," found (among other places) in the work of J. C. Scaliger and likewise familiar to Beeckman, is an example of this phenomenon.[63] But the fundamental difference between Beeckman's theory and these Aristotelian interpretations involves the former's purely geometric interpretation of the different atoms. Beeckman adhered to a strictly Democritean kind of atomism: atoms differ only in their geometric qualities. There are no inherent qualities of whatever kind: no plastic principle, no attractive or repulsive forces. Atoms are purely passive pieces of matter. During the seventeenth century, an increasing number of mechanical philosophers discovered that it was impossible to hang on to atomism without reintroducing some sort of vitalism in the form of "active principles" or life-giving "seeds" (semina). Explaining natural phenomena solely in terms of matter and motion was no longer sufficient. Even the most enthusiastic mechanists, so the historian Bruce Moran noted recently, "needed to think a little outside the box when faced with explaining the artfulness or ingenuity of nature and with comprehending the workings of the parts of that animate machine called the human body."[64] Yet by the time natural philosophers had concluded that such active principles were essential, even for the understanding of inanimate nature, Beeckman had died; in a sense, his atomism could be viewed as radical only because he articulated it so early in the seventeenth century.

Although atoms had been only a mental construction for the ancient atomists and their modern followers, Beeckman was so sure of their existence that he sketched visual representations. Pictures are not uncommon in the *Journal*; there are about 240 illustrations, ranging from very simple geometric diagrams to intricate pictures of real and imaginary machines. He even devised a perpetuum mobile—or at least a machine that looked like one, because Beeckman understood that a perpetuum mobile was impossible. Most of the *Journal* illustrations are fairly ordinary. They resemble the illustrations one finds in medical treatises and books of the time describing all kinds of machinery, textbooks of geometry, and catalogs of botanical gardens.[65] Some are very abstract, like the geometric diagrams, some are very realistic, like the drawings of machines, and some are in between, in the sense that they depict the fleeting phenomena of movement, as in the motion of an object or the path of a ray of light.

Much more interesting than these figurative and mathematical illustrations are the illustrations that one might label as "natural philosophical." Natural philosophical illustrations pretend to explain certain phenomena by picturing particles that are themselves invisible. Such pictures hardly existed before the seventeenth century. Textbooks of natural history usually lacked visual aids. Natural philosophy relied primarily on logical and semantic reasoning, deducing effects from causes and explaining natural phenomena in terms of matter and form, substance and accident, and other purely verbal, abstract concepts. By the middle of the seventeenth century this had changed dramatically, mainly because of Descartes's abundant use of pictorial devices in the *Essays* of 1637 and his *Principia philosophiae* of 1644. His explanatory concepts—like the special particles that cause magnetism—are as invisible as the concept of form that plays a central role in Aristotelian natural philosophy, but they are not abstract and can be visualized.

Given the absence of pictures in natural philosophy before Descartes, it is interesting to examine a late August 1618 note in Beeckman's *Journal* concerning his explanation of the refraction of light (see fig. 3). In this note, he represented a ray of light, not by a straight line, as was common in geometric optics, but by a chain of small circles.[66] In the figure AB represents the surface of water or a piece of glass, while DC is a ray of light. Beeckman then reviewed what happens when the light enters (or leaves) the water. He knew that the light would not continue in a straight line, but he was curious about the causes of that phenomenon. Following his corpuscular theory of matter, he viewed light as consisting of small particles called *igniculi* (more about them later). Therefore, he represented the ray of light by a chain of small circles, each little circle representing a light particle. Indeed, Beeckman intended these little circles to be a realistic

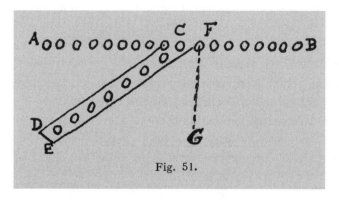

Fig. 51.

Figure 3. The refraction of light according to Beeckman
(*Journal*, 1:211)

pictorial representation of the atoms of light and not a mere symbolic representation of light (or else Beeckman might simply have used dots instead of circles). The mechanical properties of the particles were crucial to Beeckman's explanation of refraction, which is why the drawing is "realistic." According to Beeckman, the decisive factor in explaining the behavior of a beam of light was the fact that one side of a particle of light entered the other medium slightly earlier than the other side of the particle on passing the line between water and air. So this illustration was not meant simply as a fictional aid to the understanding of a phenomenon in nature but as a serious representation of what happens—on a microscopic level, we might say—when light travels from water to air. Beeckman's readiness to draw pictures of atoms in 1618, when it was still a rare phenomenon, testifies to his strong belief in the reality of the atoms.[67]

Light was only one of the many natural phenomena that Beeckman explained by invoking an atomist or at least a corpuscular mechanism. Although he postulated only four different kinds of atoms, this did not hinder him from developing explanations for the rich variety of nature and the widely diverse characteristics of matter. Four distinct atoms could be arranged in so many different ways—especially if one could make spatial configurations of them—that they could form an infinite number of substances. In 1617 Beeckman wrote that substances consisting of the same atoms exhibited divergent characteristics owing to the differing configurations of atoms (in fact, Beeckman was describing a phenomenon identical to what is now known as isomerism).[68] Elsewhere Beeckman made the analogy—found in Lucretius—to letters of the alphabet and words: although the number of letters is limited, by varying combinations of letters, sometimes the very same letters, an infinite number of words could be formed. Similarly, a limited number of atoms in varied combinations could form an infinite number of substances.[69]

According to Beeckman, therefore, problems associated with the concept of species that had occupied logicians and natural philosophers for generations could be solved elegantly. One problem was the relative stability and distinctiveness of species in nature. How could it be explained, philosophers had asked, that all kinds of differences exist among human beings—or, put another way, that between one person and another there were so many transitional forms—while there were no transitional forms between man and lion, or between gold and silver? Why were there continuous transitions in one case and discrete transitions in the other? Beeckman evidently did not agree with Lucretius and other classical atomists, who had ascribed the creation of individual species to coincidence, to a "fortuitus consursus."[70] Yet the opposite theory, the biblical story of the direct creation by God of each and every individual species, did not satisfy him either.

His solution was a compromise between Epicurus and Moses: living beings initially did not emerge by pure coincidence from the coagulation of atoms (the dysfunctional coagulations or monsters having disappeared in the meantime) but had been put together from parts that were originally created by God with the explicit intention of these parts coming together as a particular kind of living being. In 1613 or 1614, Beeckman used the analogy of the architect to explain this idea: just like the architect prepares the parts of a house (doors, windows, roofs, stones, etc.) before building it, and King Solomon prepared the building blocks of the Temple in Jerusalem, with which he later built the temple without the force of a hammer, so God fabricated the "fundamental parts" (primordia) of all things in nature in such a way that they fitted each other perfectly, as a key fits a lock.[71] He did not have to create all the individual species, but he could create a large number of living beings with a limited number of fundamental parts, just like human beings can create a large number of syllables, words, and sentences with a limited number of letters. Thus, Beeckman integrated the theory of divine creation with the idea that nature always works in the same way, which is a prerequisite for natural philosophy. Once God has created the fundamental parts, these parts cannot but bring forth the creatures God had intended them to bring forth, as Beeckman explicitly stated in 1620: these "principia," once created by God, "could not but form a specific being" (non possint non hoc facere).[72] God directly created only man, because God had to add "particles of a divine nature" (particula divinae aurae). Also, God did not leave the creation of large animals to the "concursus" of the preestablished parts because the chance that all these parts would come together in the foreseeable future was simply too remote.[73] But small animals and the rest of nature were indeed created through the fortuitous coming together of the preordained parts.

Somewhat later Beeckman gave a more detailed illustration of how we should conceptualize the primordial: he used the image of the regular geometric polyhedrons in September 1620. One can build different Platonic polyhedrons with a specific number of small equilateral triangles (4, 8, and 20), and objects that are built of octahedrons form a different species compared to objects constructed from icosahedrons. An intermediate number of triangles cannot form a polyhedron, and, according to Beeckman, that explains why there are no intermediate species between, say, man and dog. Certain variations appear within a species—such as between Achilles and other Greeks or between different varieties of dogs—on account of small variations in the constituent parts, the primordia. But there are unbridgeable differences between species, such as the difference between man and dog, or silver and gold.[74]

Thus, Beeckman placed a new analytical level between visible matter and the postulated atoms that could, in principle, make atomism physically and chemically fruitful. All known substances consist of small particles that are characteristic of that substance. These substance-specific particles were themselves composed of atoms. Just as language is not composed of separate letters but of words, and just as words are the smallest meaningful units in any specific language, so it is not the independent atoms but the species-specific combinations of atoms that constitute the lowest level of analysis. Beeckman used various words to describe these substance-specific combinations: at first he called them "primordia"; later he used the word "infima species"; and finally, in August 1620, he coined the term "homogeneum" (later expanded to "homogeneum physicum").[75] Thus, historian Henk Kubbinga credits Beeckman with the first formulation of the notion of the molecule (a term Gassendi introduced in 1637, but which gained currency only in the nineteenth century). Whether or not this claim is valid, Beeckman in principle made an important contribution to bridging the gulf that separated ancient atomism from modern theories of matter by introducing these intermediate particles. But the direct impact of his ideas was minimal. Furthermore, Beeckman was more interested in physics than in chemistry, and thus more interested in atoms than in homogenea—or whatever he called these intermediate particles.[76]

Not long after the defense of his dissertation, Beeckman began to see that the evolution of his early corpuscular theory of matter into a form of atomism resulted in a contradiction; and in the course of 1620, he realized that atomism had an inherent problem. By definition, atoms were absolutely hard, but in order to bounce off other atoms after collision, they had to be elastic as well—and these two qualities were mutually exclusive.[77] He initially viewed the problem as a warning against seeking first principles that were too specific. Moreover, the difficulty arose only when the atoms were set in motion: as long as he dealt with the composition of matter (e.g., chemistry and pharmaceutics), there were no problems.[78] But the movement of atoms was essential to Beeckman's entire natural philosophy, even more important than their form, and therefore he continued searching for solutions. He repeatedly sought an answer in the composite character of atoms, which he at first had regarded as indivisible. Previously, he had accepted that air particles do not float about randomly but form segmented rings—which can be compressed temporarily like hoops and then spring back to their original position when pressure is removed—in order to explain the elasticity of air.[79] He also had compared air to a sponge.[80] After 1620, Beeckman postulated that atoms themselves resembled a sponge. What initially had been compact, massive particles of matter then became malleable, sponge-like bits, which contained pores and holes so small that only the very smallest particles of the element of fire, igniculi, could pass through them.[81] These igniculi were present everywhere in the world space, like a sort of ether, but they did not fill it entirely. By the compression or bending of atoms and combinations of atoms, they were temporarily squeezed out; but by their rapid and continuous movement over a period of time, they nonetheless again penetrated the atoms, so that they returned to their former shape.[82]

It is questionable whether Beeckman was satisfied with this makeshift solution. He simply transferred the problem to the subatomic level. Beeckman expressed doubts about the viability of atomism; hence, it is understandable that his brother Abraham did not publish those *Journal* entries with a very specific development of Beeckman's atomism—for example, those from 1620 of the "homogenea physica"—after Isaac's death.[83]

VACUUM AND AIR PRESSURE

Whether Beeckman was an atomist or simply believed in the existence of unspecified corpuscles, he always advanced the idea that there is empty space between the particles of matter. Beeckman had to find a new explanation for a series of phenomena that had been explained since antiquity by a natural tendency to prevent the formation of a vacuum at all costs. In the concept of air pressure, Beeckman found the replacement for the false concept of horror vacui or "fuga vacui."[84]

One of the most important arguments against the possibility of a vacuum had always been that a vertical tube sealed on one side could draw water from a pool to a very high level. Apparently, the inclination of nature to prevent a vacuum was so strong that it could overcome even the natural tendency of water to fall downward. The effect of the horror vacui could also be observed in a pump drawing water upward. Beeckman, who had been able to study this phenomenon many times as a craftsman, took a different view. As one of the corollaria accompanying his 1618 thesis stated, "Water that is sucked up by a pump is not drawn by the force of the vacuum, but rather pushed up by the pressure of the air."[85]

Beeckman deliberately spoke about "the pressure of the air" and not about its weight or heaviness. As he formulated the explanation at the time, the water rose because of the "pressure of the air above" (propter pressionem incumbentis.)[86] The cause of this phenomenon is the tension or "co-actus" of the lowest layers of air, a tension that is itself caused by the weight of the column of air above it. Thus, the weight of the air is only indirectly responsible for the rise of the water. This is significant, because in this way it is possible to explain why water also rises in a suction pump when a cover is placed on the pit, which in principle neutralizes the weight of the air column above the well. According to Beeckman, the lowest layers of air were already in a compressed state before the closing of the pit, which thus did not change that condition. The same applies to a vessel that is completely sealed from the outside air, although the effect of air pressure in that case is more limited. If a quantity of air is sucked out of such a vessel, the tension and therefore the pressure of the remaining air are markedly reduced. It then becomes more difficult to suck out still more air.[87]

Beeckman's explanations were incomprehensible or absurd to most of his contemporaries. According to some interpretations of Aristotelian philosophy, air had its own natural place and by itself had no weight: therefore, it could exert no pressure. Everyday experience appeared to confirm this statement. Although man is surrounded on all sides by air, he never feels the pressure of that air. Beeckman disagreed. In his opinion, elements always had weight: the sensation of lightness (levitas), existed only because of the difference in weight between air and water.[88] The reason one feels nothing of the pressure of the surrounding air in everyday life is because air exerts equal pressure on all parts of the body, including from below and from the sides: one feels only differences in pressure. As long as there are no differences in pressure, one is not able to notice air pressure, just like a fish is not able to sense the pressure of the surrounding water. Beeckman took this argument from Simon Stevin.[89]

Beeckman initially was enthusiastic about the explanatory possibilities of the concept of air pressure. Whenever attractive forces came into play, he showed how air pressure explained the phenomena. The fact that a burning candle sucks up its own wax was relatively straightforward. The burning of the fatty parts in the wick creates a vacuum, which is quickly filled by the contiguous, already melted fat particles, which are driven in the direction of the wick by air pressure.[90] Somewhat more complex was the explanation of cohesion: why does a stone remain intact, while a pile of sand readily falls apart? According to Beeckman this was because all of the particles of matter of the stone fitted together perfectly, leaving no spaces in between, so that the air pressure could operate in only one direction, from the outside in. Thus, the particles remained pressed hard together. The cohesion of particles became greater as the stone became smaller: smaller objects have a relatively great surface area, and air pressure is exerted precisely on the surface. With this combination of a physical principle (air pressure) and the mathematics of isoperimetric figures, Beeckman thought he had produced a simpler and more comprehensible explanation of cohesion than had the German chemist Andreas Libavius, for example, who was well known to him as the author of *Alchemia*, published in 1597. Libavius had proposed a "plastic principle" (plasticum principium), created in nature by God.[91]

Beeckman's explanation of magnetism was more complex. Magnetism was not just another phenomenon but one that posed a challenge to any mechanical interpretation of nature. A mechanical philosophy usually has no problems in reinterpreting the "manifest" qualities in nature (like color and sound), but the "occult" qualities—such as chemical affinities, electrical phenomena, and magnetism—were harder to fit into a mechanistic program, because one had to incorporate concepts of action at a distance into one's theory.[92] Beeckman was well aware of the extraordinary significance of magnetism: "How the lodestone [in Dutch, seylsteen] attracts iron to itself is the most wonderful matter that I know of the things that one can see happening with one's own eyes, and I believe that the correct explanation of it would be very advantageous to us in the discovery of many phenomena."[93] Thus, if he could use his natural philosophy to give a good explanation of magnetism, this would definitely strengthen his philosophy. He could not base his explanation, however, on the existence of attractive forces working at a distance: "The attraction in my judgment is not according to nature . . . since in nature nothing happens without contact."[94] For this reason, Beeckman later rejected the concepts of William Gilbert, whose book *De magnete* he first read in 1627.[95] Gilbert theorized that the attractive force of a terrestrial magnet was not of a material nature. The Earth was endowed with some sort of intelligence, which reacts to the nonmaterial magnetic forces of the sun and the stars. According to Beeckman, this was an explanation unworthy of a philosopher. A good explanation must rest exclusively on the effective contact of particles of matter.[96]

Beeckman already had conceived of a mechanism that caused magnetic phenomena in 1614. He hypothesized that a magnet sends out a very fine material, the "spiritus," in all directions. The particles of this material can penetrate the pores of certain bodies, in particular iron. The form of these pores in the iron object corresponds exactly with that of the particles of the spiritus. At those points where the spiritus of a magnet penetrates the interior of an iron object, the pressure of the surrounding air is lifted, while the spiritus particles, because of their perfect form, exert no pressure on the iron. On the other side of the iron object, however, the air pressure is not reduced. As the side facing the magnet experiences less or no air pressure, whereas normal air pressure is exerted on the other side of the object, the object is pushed in the direction of the magnet.[97]

Beeckman's explanation seems largely inspired by Lucretius's writings on magnetism. Lucretius, however, had speculated that the iron object was pushed to the magnet like a ship is pushed forward by the wind.[98] The concept of low pressure is absent in *De rerum natura*. Beeckman's explanation, however, fits perfectly with his theory that air, like water, is subject to the laws of hydraulics. I do not believe it is a coincidence that, in the note in which Beeckman argues against the "fuga vacui," he mentions Simon Stevin's *Van de Weechkonst*. However, whatever the inspiration for his theory, it certainly is a highly ingenious explanation that raises all sorts of questions: If the spiritus particles come out of the iron object on the back side, would not the air pressure be reduced there too? And what about the form of the particles that make up the spiritus? At first, Beeckman adapted his explanation from Lucretius's, asserting that the particles had little barbed hooks, one dragging along the other. Later he settled on a conical form, with a sharp point and a broad base.[99]

Beeckman also believed he could use his air pressure theory to explain other phenomena generally regarded as puzzling: the efficacy of purgative drugs, electrical attraction, and especially the motion of the tides. Although the connection between the movements of the moon and the tides was widely accepted, there were great differences of opinion regarding how the moon exerted its influence. Stevin, in his *Wisconstighe Ghedachtenissen* (Mathematical Memoirs, 1608), wrote vaguely of a sucking force, but then he limited himself to a purely mathematical description of tidal movement.[100] Beeckman, who had read Stevin's work during his student days in Leiden, found this explanation wanting: there also had to be a physical explanation.[101]

Initially, Beeckman theorized that a mechanism, analogous to the mechanism that caused magnetism and rested on the principle of pressure, connected the moon and tides. Beeckman thought the moon continuously emitted particles that could rarefy the air in two places on Earth: directly under the moon and also on the far side of the Earth, where indirect radiation created an "opposition point," a "virtual moon," one might say. Air pressure diminished in these two places. Thus, seawater was pushed into the two low-pressure areas, causing a flood tide there and an ebb tide in the normal air pressure areas.[102] In 1623 Beeckman considered an alternative theory, whereby the particles streaming from the moon caused a slight expansion of the water right under the moon, not a rarefaction of the air. In Beeckman's revised theory, differences in air pressure no longer play a role.[103] In 1630, when Colvius showed Beeckman a paper on the tides written by Galileo in 1616—the *Discorso sopra il flusso e reflusso del mare*—Beeckman's ideas about ebb and flood shifted again. In this treatise, which circulated throughout Europe in manuscript form and would later be reworked in the *Dialogo*, Galileo attributed the movements of ebb and flood to the combined effects of the daily and yearly rotation of the Earth. This was an attractive and purely mechanistic theory, which did not require speculations about small and invisible particles, and Beeckman endorsed it.[104]

Beeckman went even further than Galileo. He thought that perhaps the movements of the air and the clouds could be explained in the same way that Galileo had explained the movements of the oceans. Thus, by extending Galileo's astronomical theory of the tides to meteorological phenomena, Beeckman hoped to place his long-cherished interest in the weather on a solid mechanistic foundation. And could he not explain the moon's movement in a similar manner? Beeckman conceived of the moon as an object that floated on the highest layer of air and was thus subject to movements in the air layers.[105] While Galileo had eliminated the link between the moon and the tides, Beeckman reintroduced the moon's movements as resulting from the same causes that produced the tides. After 1630, however, nothing more is heard from Beeckman about a connection between air pressure and tides.[106]

In the meantime, perhaps on reading Gilbert's *De magnete*, perhaps slightly earlier, Beeckman had given up his air pressure theory. He realized at some point that his explanation of magnetism was incorrect. When a closed cistern, filled with air, separated the magnet and the piece of iron, the under-pressure mechanism did not work, and yet the magnet attracted the iron. Also, when solid glass separated the magnet and the iron, the air pressure could not operate as Beeckman had theorized, yet the magnetism phenomenon was still present. In the absence of air, air pressure cannot give way to spiritus particles; hence, magnetism cannot be attributed to changes in air pressure. Beeckman now explained magnetism using the flow of spiritus or world ether, which seemed to him a more general theory: "Universalior igitur ratio videtur reddi per hunc ignem quam per aerem."[107] He also rejected his explanation of cohesion: he understood that it was highly unlikely that a stone would simply fall apart in a vacuum. And what would happen when a heated stone, in which the back-and-forth movement of shooting fire particles might cause disintegration, would be placed in what Beeckman called "the Epicurean interworld" (in intermundiis Epicuri)? Most likely it would not crumble because of the absence of air.[108] Beeckman initially hoped to save his explanations by assuming that the particles of stones, for example, were shaped like little hooks, but he did not believe that this was a satisfactory solution to the problem.[109] He came upon a real solution only when he replaced the pressure of air particles with the fire particles or igniculi, which were also present in a vacuum.[110] The pressure of igniculi in magnetism and in cohesion assumed the role of air pressure, without actually changing the mechanism.

THE WORLD ETHER

Beeckman's use of fire particles for his explanations of several phenomena may be viewed as the introduction of an ad hoc hypothesis, which enabled him to stick to his corpuscular notions. But it may also be viewed as Beeckman's attempt, perhaps born of necessity, to provide a more general explanation of natural phenomena. Beeckman at least saw it that way. After he had explained magnetism by the pressure of igniculi instead of air pressure, he wrote, as we just saw, that with these fire particles he had given a more *universal* explanation of the phenomenon.[111] Beeckman's ideas about the structure of the world clearly evolved in the direction of an ether theory, in which the world ether—an omnipresent, extremely fine, and universally permeable substance—is the dominant element in a sun-centered universe.[112]

Beeckman first mentioned the igniculi in 1613 upon exploring the question of heaviness.[113] He did not want to assume that this quality was inherent in matter. Heaviness must be explained by the form, size, and movement of small particles of matter, as the effect of an invisible material mechanism behind the phenomena, just as in the case of color and taste. First he considered whether heaviness was perhaps the effect of a whirling movement. Beeckman frequently had witnessed in the harbors of Veere and other towns in Zeeland how great objects in a whirlpool (vortex) were drawn toward its center. But he rejected this hypothesis right away and proposed instead that the explanation must be sought in the effects of a "continual, downward stream of fire particles" (defluxus subtilium corporum), which originates in the sphere of the fixed stars and is directed toward the center of the world, from which the stream is sent back again to the stars. This shooting of corpuscles back and forth is so fine that it can pass through all material objects and exert a force on the external as well as the internal parts of objects. Thus, earthly objects are not subject to the force of the stream of subtle matter in proportion to their surface area, as with air pressure, but in proportion to their volume or quantity of matter.[114] In other words, heaviness is the sum of the pressure of subtle matter on the constituent parts of earthly objects.[115]

The sun functions in the center of the world as the focus of the rays that are sent down from the starry sphere. After 1616, Beeckman took for granted the fact that the sun was at the center of the universe. He had hesitated to do so during his student years at Leiden. He understood the arguments in favor of the Copernican system and seemed to subscribe to them (he read Stevin's defense of Copernicus at that time), but at other moments he still reasoned as if the Earth was at the center of the universe.[116] But Beeckman changed his mind in 1616, possibly under the influence of Lansbergen. After pondering the behavior of the light particles in the universe, Beeckman found it much easier to assume that the sun was in the middle of the universe. The rays of light coming from the stars are concentrated in the sun and partially reemitted in the form of sunlight, in the opposite direction.[117] Beeckman makes an analogy with a burning candle, which draws up its own fuel and emits it as light.[118] This radiation also reaches the Earth and has a remarkable effect. An earthly object must feel more force during the daytime—because sunlight does not actually differ from stellar radiation—than at night, when only stellar radiation is at work. Therefore, the object is heavier during the day. This contradicts everyday experience, but Beeckman does not immediately dismiss the idea because on the barge from Utrecht to Amsterdam during a night voyage, he heard the captain say that it was easier to move a vessel at night than in daylight.[119] Beeckman granted that the difference in the intensity of radiation at night and in the daytime is extremely small. The intensity of sunlight appears to be greater than that of the subtle matter emitted directly by the stars, but that only seems so. Stellar radiation is barely observable with the naked eye, because stars emit only faint light. Their radiation becomes visible in the form of light only when in passes a certain limit, so Beeckman speculated. The large quantity of stellar radiation already present can suddenly be transformed into clear light through the addition of a small quantity of sunlight: sunbeams are like the proverbial straw that breaks the camel's back.

Initially, Beeckman declared that heaviness resulted from the driving force of stellar matter. Later, he posited that the heaviness effect resulted from the reflection of a part of the stellar matter by the surface of the Earth. The reflected particles of subtle matter are first mixed with coarser earthly matter, and the coarser matter is then shaken off by the great speed of the particles of subtle matter. Radiation becomes so pure by the time it reaches the highest layers of air that it ignites spontaneously. Particles of stellar matter then radiate in all directions, as well as back down to Earth. According to Beeckman, the effect of heaviness occurs at that point, and not sooner.[120] He realized that we on Earth see nothing of the burning stellar material in the highest layers of the air, but he maintained that, if we could place ourselves at a great distance from the Earth, we would observe it as a great light-emitting body, comparable to the sun. Perhaps, Beeckman speculated further, what we see as the sun is not the real sun, but the sphere of fire around the "core-sun," exactly as with the Earth.[121] Sometimes Beeckman evidently lost control of his speculative inclinations.

Beeckman was not always clear about the precise relationship between fire, light, and subtle stellar matter. Light and fire are sometimes identical; sometimes two independent effects of the same cause; and sometimes one emerges from the other. In general, Beeckman used the concept of "fire" (ignis) in two distinct senses: first, as the visible, relatively coarse element of fire; and, second, as the smallest particles from which the coarse fire is constructed, the igniculi or fire particles. These igniculi can join together to form small piles and then form a particle of light, which may be observed as such only if it attains a certain speed. Light, Beeckman proposed, is rarefied fire in the sense of the coarser element fire, and thickened fire in the sense of fire particles.[122] Light particles are thus constructed of fire particles, have a composite nature, and consequently are not atoms—although Beeckman was sometimes so careless as to speak of "light atoms."[123]

Beeckman theorized that light had a corpuscular character and was not an effect of a change in the condition of the medium between the eye and the light source. He strongly rejected the Aristotelian position that such a change in the medium produced "species visibiles," the images of the light-producing object, in one of the corollaries attached to his dissertation in 1618: "What the writers on optics call species visibiles are in reality material bodies."[124] He also later distanced himself from Kepler's medium theory and rejected Descartes's hypothesis that light is the result of the propagation of pressure or of the inclination to movement in a medium.[125]

Beeckman used various arguments to support his emission theory. If light existed as a result of change in the medium, then light could not exist in a vacuum. According to Beeckman, outer space was effectively empty, and yet light still reached the Earth from the sun and the stars—which thus could not be the effect of a medium. Beeckman also argued that beams of light do not extinguish each other. Such a phenomenon must occur, according to Beeckman, if light were caused by changes in the medium, because the medium could not possibly change in two different ways at the same point. Yet nothing is more intelligible than the hypothesis of two streams of particles, crossing and splitting each other.[126] Thus Beeckman also disputed the analogy made between the propagation of light and the propagation of water waves, where propagation occurs by means of a change in the medium. Mutual extinguishing or at least disturbance occurs with cross currents of water waves. Finally, he dismissed the comparison between water and sound waves; sound waves likewise resulted from moving particles.[127]

Beeckman had to accept that light was not propagated with infinite speed as a consequence of his emission theory; a particle cannot be in two places at the same time. He indeed regularly opposed the view that light could be propagated instantaneously, and he was even prepared to carry out an experiment to prove he was right when he met Descartes, who held the opposing position.[128] Behind a wall a quantity of gunpowder was exploded. While observers on the other side of the wall heard the sound of the explosion immediately, the flash of light reached them only indirectly, through its reflection in a mirror located far away. Now the question was: Which reached them first—the sound or the light of the blast? Descartes and Beeckman did not notice any difference, but there were so many objections to the experiment that the result could not be determined conclusively.[129]

Beeckman clarified a large number of previously discussed topics with the aid of his igniculi. He attributed the appearance of colors to the surface structure of objects, which absorbed some light particles while reflecting light particles of a somewhat different shape.[130] Also, he used the movement of igniculi to explain cohesion, magnetism, the movement of the tides, and air pressure. Air is heavy because fire particles from outer space are continually colliding with it. Compressed air expands because igniculi try to separate the air particles and push them apart by continuous movement. Just as a gladiator keeps his opponents at a distance by swinging his sword, a fast-moving fire particle drives the air particles apart.[131]

Other phenomena found their place in Beeckman's natural philosophy thanks to the igniculi concept. He attributed seawater's salinity to the fact that starlight and sunlight are precipitated as salt when they fall on the sea. Salt was strongly concentrated fire come to rest, "ignis coactus" or "ignis condensatus."[132] According to Beeckman, the fact that sea salt and seawater produced light under certain conditions confirmed his theory. Heat was also an effect of igniculi, because it consists of the active movement of constituent particles of an object, motion caused by the igniculi in and around the object.[133] At times, Beeckman appeared to attribute heat to a separate substance, "caloric matter," which differs qualitatively from ordinary matter, but this is incorrect.[134] Beeckman did not always express himself carefully, but elsewhere he made it abundantly clear that those "heat particles" are actually ordinary fire particles. Cold was simply the absence of heat: the difference between a "lack of heat" (absentia caloris) and "positive cold" (frigus positivum) that he originally made in good Aristotelian fashion later disappears from his notes.[135] Beeckman attributes the fact that water expands when freezing simply to escaping fire particles pushing the water particles apart, and not, as the standard theory had it, to the water's taking up particles of cold.[136] Finally, Beeckman explained the flammability of certain materials as a phenomenon caused by fire particles. As noted, these fire particles could be present in concentrated form in salt but also, and primarily, in oils and fats.[137] In fact, according to Beeckman, oil and fat were nothing more than the coagulations of fire particles that had lost their motility.[138] During combustion, the oil particles become separated from each other and rarefied; they recover their former mobility and are emitted as light and fire. It is probably not a coincidence that fire particles play the leading role in the natural philosophy of a candlemaker.

Yet it was not only phenomena from everyday life that Beeckman ascribed to the motion of fire particles. He thought he could also explain the structure of the solar system and the motion of the separate heavenly bodies with his ether theory (once he had accepted the central position of the sun, Beeckman never doubted the Copernican system).[139] In late 1628, after reading several works by Kepler, Beeckman realized that his own igniculi actually offered a splendid explanation of various heavenly phenomena, which he should perhaps publish.[140] He had studied these phenomena earlier. He initially believed that the planets maintained their original God-given motions around the sun because there was nothing to stop them—the source, as we shall see, of his principle of inertia. Later, however, he tried to devise a corpuscular mechanism. He imagined that oblique pores extended out of the sun, through which subtle matter was emitted, which made the sun revolve on its own axis. Beeckman compared this to the primitive steam engine described by Hero of Alexandria. This engine consisted of a hollow sphere filled with hot vapor that escaped through obliquely placed pipes, thus making the sphere turn on its own axis. By means of the corpuscular radiation it emitted, the turning sun then also gave the planets a certain motion, one on their own axes and one around the sun, a drag effect, as it were. This resulted in the daily and yearly motion of the Earth and the motion of other planets.[141] Beeckman thought that even what is called the third move-ment of the earth—the trepidation and the precession of the equinoxes—could be explained with the help of the radiation of subtle matter.[142] And he was very satisfied with his explanation, as indicated by this 1628 note: "By these means I have shown that all three earthly movements are performed without any fictive internal force and that they follow in a mathematical way from the movement of particles that are emitted by the sun."[143]

The consistency of Beeckman's ideas pose several problems. Was Beeckman still convinced of the existence of empty space around 1630, given the signifi-cant role that he had assigned to the igniculi and the subtle matter emitted by the sun and the stars? It appears at times that Beeckman held to an unstated assumption that outer space was largely filled with subtle matter, which would be the corpuscular counterpart of the Aristotelian ether, the fifth element. Was there indeed such a great difference between Descartes's vortices of "matière subtile," which filled the entire universe, and Beeckman's "defluxus subtilium corporum," which played such a dominant role in his world system? Further-more, we should note the similarities to the Stoic philosophers' "pneuma" that filled the world. When Beeckman described his subtle matter as "materia aeth-erea" or "substantia energetica," he seemed to place himself in the tradition of late Renaissance thinkers such as Bernardino Telesio, Giordano Bruno, and others who also regarded the world system as filled with such substances.[144] But with Beeckman, there can be no question of a completely filled space. His light particles also are composed of smaller atoms, and "interstitia inter ignicu-los vacua" exist between these atoms.[145] Perhaps Beeckman's world was more crowded in 1628 than in 1618, but it was certainly not completely full, as was the case with the Stoics and Descartes.

Beeckman's regular comparisons of the sun to a child's spinning top also poses a problem.[146] This toy was a small-scale model of the heavenly bodies for Beeckman, just as the magnet had been a model for Gilbert. The top's extraor-dinary properties entailed the fact it could raise itself by its own motion and also simultaneously spin on its own axis and have circular motion on the floor. Beeckman closely observed the actual behavior of tops and did not content himself with speculation in his study. He sometimes invited children who were outside playing with a top to come to his study to carry out all sorts of experi-ments. Thus, he once let a spinning top land in water in order to see what effect the surrounding medium would have on the motion of the top.[147] Such experi-mentation did not produce clear results, but it is noteworthy that here Beeck-man tried to explain the motion of heavenly bodies by beginning with the mechanics of moving objects in general, without immediately resorting to a corpuscular mechanism. It remains unclear how this more mathematical and descriptive approach to the motion of the heavenly bodies would be consistent with the more corpuscular and explanatory approach he favored in other notes.[148]

CHAPTER FIVE

Principles of Mechanical Philosophy II: Motion

Given that the mechanical philosophy begins with the assumption that matter is purely passive, what then is the cause of change in matter? Change cannot be the result of any inherent tendency in matter to change, which would contradict the assumption that there are no inherent qualities or powers in matter. Change always arises from external forces that compel matter to change. Such forces, in turn, emanate from other purely passive matter. Now, this phenomenon can only occur through motion. Change in matter always results from motion. A theory of motion therefore is an indispensable ingredient of any mechanical philosophy. Atomism, corpuscularism, or any other theory of matter in itself does not imply such a philosophy. Beeckman's philosophy of nature thus requires a theory of motion to be a *mechanical* philosophy of nature. And, indeed, Beeckman's understanding of motion no longer proceeds from Aristotle's qualitative, hierarchically ordered cosmos—with its inherent tendencies and natural places of which matter in some mysterious way seems to be aware—but from one that coheres with Beeckman's conception of inert matter.

THE PRINCIPLE OF INERTIA

In traditional Aristotelian natural philosophy, the concept of motion encompassed a wide variety of phenomena, including local motion or displacement as well as the growth of a seedling into a mature tree and the decomposition of a pile of leaves into humus.[1] All development and becoming, all coming into being and decay were called motion. Motion thus was a process, a physical change in an object.

The phenomenon of local motion, the kind of motion treated by mechanics, could be interpreted in terms of growth processes just as easily as phenomena in natural history. The starting point for this view is that the world is a hierarchically organized whole, an order in which each object has its own natural place and its own goal toward which it strives. The place where an object finds itself is thus not merely an external coincidence, as if it would make no difference at all where an object is located, so that object could just as well be somewhere else. Any object that is not in its *natural* place has not yet reached its ideal situation, and it is therefore still incomplete. Yet it does show a desire to repair this situation by taking up its natural place. This then leads to local motion. Once the object reaches its natural place, the object comes to rest. Rest is the norm; motion, a deviation from it.

Motion requires an explanation *because* it is not normal. This view leads to the thesis, in Aristotelian philosophy, that every motion must be caused by a mover.[2] When this cause of the motion ceases to operate, the motion also stops. Everyday experience confirms this: a wheelbarrow that is no longer pushed forward comes to a stop. There is a problem, however, with motions that continue without any apparent mover, such as the phenomena of falling motion and throwing motion. How does it happen that a stone, once thrown, maintains its motion after it leaves the thrower's hand? From the fourteenth century onward, a popular explanation entailed the stone receiving an "impetus" from the throwing hand, a force impressed onto the stone that acts as an internal source of motion of limited duration. By moving the stone, the agent creates a force that continues to act when the stone has been separated from the primary agent. Thus, scholars maintained the validity of Aristotelian teachings, even in situations that at first glance might contradict them.

From the outset, Beeckman found this impetus theory incomprehensible. "What the philosophers say about a force being impressed upon the stone," so he wrote in late 1613 or early 1614, "seems to be without reason. Who can conceive what this could be, or how it keeps the stone in motion, or in what part of the stone it is located?" He therefore articulated an alternative hypothesis. It is quite simple, he said, "to imagine that motion in a vacuum never comes to rest because no cause that could change the motion occurs. For nothing changes without some cause of change."[3] Apparently, Beeckman had to ascribe an almost thing-like character to impetus in order to be able to form an idea of it. For him, the world consisted of concrete, almost tangible things, and abstract concepts were intelligible only if they can be pictured as if they were things. This is a point to remember.

Beeckman's words imply a total break with the fundamentals of Aristotelian teachings about motion. Beeckman does not provide a new solution to the old problem of what keeps a thrown stone in motion; instead, he responds with an entirely different question: Why should an object moving in a vacuum come to a stop at all? He no longer assumes that every motion is in need of an explanation; in his view, only a *change* in motion requires an explanation. Beeckman no longer considers motion to be a *process* but views it as a *state*, and thus he sees no need to look for an explanation for its continuous existence anymore than any other state's or quality's existence. Once an object has been put into motion, it cannot deviate from that motion on its own. This is exactly the point he made in one of the corollaries to his 1618 dissertation: "A stone that is thrown away by the hand does not remain in motion by some added force, or because of the fear of a vacuum, but because it is not able not to continue [non potest non perseverare] the motion that it had when it was still being moved by the hand."[4] Only an external force can hinder the motion, for example, the action of the medium or an encounter with another object. Shortly after 1613, Beeckman could formulate a new principle of inertia concisely: "That which is once stirred always stirs, as long as it is not prevented from doing so."[5]

Beeckman's introduction of a new principle of inertia and his rejection of the traditional concept of impetus are not inconsistent with his continued use of the word *impetus*. In these instances, impetus was no longer the *cause* of the continuous motion but the force that is the *effect* of motion. The creation of a completely force-free universe did not arise from the rejection of the concept of impetus. That such a force of motion existed followed from the fact that a body in motion had the ability to move another object, just as it was necessary to apply a force in order to stop a moving object. The measure of this force or impetus depends on the quantity of matter (*corporeitas*) and the velocity (*velocitas*) of the body.[6] Force, therefore, remained an important concept in Beeckman's natural philosophy (as it would be in the philosophy of Descartes).[7]

Beeckman's notion of inertia had several important consequences. For instance, it seemed to allow for the possibility of a perpetuum mobile. It is not difficult to disprove a perpetuum mobile in an Aristotelian context: a constant motion requires a constant moving force. Beeckman's new principle of inertia, on the other hand, seems to suggest that a perpetuum mobile is possible after all; one must simply eliminate all impediments to movement so that, once a machine has been set into motion, it will go on indefinitely. Indeed, in a frictionless world this would certainly be the case. But the real world is not frictionless, and therefore a perpetuum mobile is still impossible to construct. Beeckman formulated his principle of inertia for motion in a vacuum and for a situation in which no friction occurs; but he also knew that anyone who claimed to have constructed a perpetuum mobile was deluding himself because he understood that such a situation does not occur in the real world.[8]

Beeckman's conception of inertia had serious consequences for the science of motion (i.e., mechanics). It implies that a one-time force exercised upon an object leads to a *permanent* change in the object's state: in principle, a one-time push against an object leads to an endless motion of that object. If, however, the execution of that force is not a singular event but is repeated continually, then the motion does not remain constant but continually increases. The first increase in motion is retained, and then the second is added to it, then the third, et cetera. While in Aristotelian philosophy, a constant force leads to a constant motion, a constant force in the world as conceived by Beeckman leads to a constant *acceleration*. This had enormous consequences for the analysis of the behavior of moving bodies.

Beeckman's concept of motion also had consequences for the idea of the cosmos as a whole. The idea of a hierarchically ordered cosmos, where each object has its natural place and where each motion stops by itself when an object arrives at its natural resting place, is completely undermined by motions that in principle continue endlessly. Under such a theory, it is no longer possible to call any one place more natural than another, and the concept of a natural place and a natural order of space thereby becomes devoid of meaning: the state of an object is no longer dependent on the place that it occupies in world space, and there is no longer any privileged, natural direction of motion. An object does not have any desire to move toward or away from the center.[9] Thus, the object is completely indifferent to the place that it occupies and the motion that it executes. And, contrary to Aristotelian doctrine, an object can simultaneously carry out several motions that do not reinforce or interfere with each other.

Naturally, Beeckman did not realize immediately all the consequences of his position. Only over the course of years did he understand that—given his new concept of motion—he had bid farewell to the Aristotelian understanding of place and the idea of a hierarchically ordered cosmos. Beeckman did not connect the concept of conservation of a one-time moving force to the conservation of a one-time received *velocity* until 1631, perhaps as a result of a conversation with Descartes.[10] Earlier he found such precision unnecessary. Nor, until a

1630 letter to Mersenne, did Beeckman explicitly assert that rest was just as natural a state for an object as motion.[11]

An important feature of Beeckman's principle of inertia (thus one can call his theory of the conservation of a one-time imparted motion) was that he applied it to circular motion as well as to rectilinear motion. In December 1618, he wrote:

> In a vacuum everything that has been moved once remains in motion always, either following a straight line, or in a circle, both around its own center (as the daily rotation of the Earth) and around [another] center (such as the annual revolution [of the Earth around the sun]). Because the smallest part of the orbit is curved, and in the same manner as the entire orbit, there is no reason why the circular annual motion of the Earth should deviate from that curved line and change to a straight one, because a straight line is not more natural than a circle and is equal to it in nature and extension. For a part of the circumference relates to the whole in the same manner as a part of a straight line relates to the straight line as a whole.[12]

Naturally, Beeckman understood that rectilinear motion generally prevails over circular motion in the world around us. He knew that a stone set loosely on top of a horizontally rotating wheel would fly off that wheel. But he attributed this to the effect of the medium, as will be shown later on.

Beeckman's principle of inertia differs from the Cartesian or Newtonian principle of inertia, which has become one of the foundations of modern mechanics, because it applies to circular motion as well as to rectilinear motion. Yet this is no reason to deny that Beeckman's concept of inertia is essentially modern. Beeckman had taken the crucial step in asserting that only *change* in motion must be explained. The complete dismantling of the Aristotelian cosmos follows from this basic principle, regardless of ensuing and significant changes in the application of this principle of inertia.[13]

How did Beeckman free himself from the Aristotelian idea of motion? Was the principle of inertia something that became clear to him in an instant, as an insight that suddenly broke through and that immediately placed everything in a new light, or did it develop gradually? Beeckman's 1612 note on the motion of heavenly bodies—a year before his first formulation of the principle—contains a revealing passage. In response to a remark in Scaliger's *Exotericarum exercitationum de Subtilitate H. Cardani*, Beeckman wrote: "It appears to me that a heavenly sphere [coelum] is not moved by Intelligences or by the continuing action of God, but because, once it has been moved according to itself and the nature of the place [sua et situs natura], it can never come to rest by itself. What can be done with a few means, is said to have been done badly with many."[14] Thus, Beeckman arrived at his principle of inertia in an astronomical context when he proposed that a moving sphere, or any moving object, never can come to rest on its own. He did not arrive at this position by observation of the phenomena around him (because the impediment of the medium makes it impossible to observe the effect of inertia), but rather through reflection on the motion of heavenly bodies, an extremely abstract topic. Thus, Beeckman's principle is not an abstraction from observed phenomena, but an "observable" phenomenon in an abstract world.

Beeckman seems to introduce an important limitation in the quoted passage (as Galileo did shortly before him): the principle applies only to an object that is in its natural place and stays there. This represents a marginal case, which allowed Beeckman at first to remain within the Aristotelian framework. Aristotelians took it for granted that the motion of the heavenly spheres was circular and perpetual. The question was only how that perpetual motion was sustained. Beeckman had an unconventional answer, with unforeseen implications: he said that an explanation was not necessary in these exceptional circumstances. Later, Beeckman realized that the fundamental concept behind this answer probably had broader significance and could be applied to other motions, causing the originally accepted idea of natural place to lose its meaning. Thus, Beeckman's formulation of the principle of inertia in his dissertation consisted of at least two stages. First, he applied it only to a particular case, the heavenly spheres that, while revolving, always remained in their natural place. In doing so, Beeckman remained within the Aristotelian framework. Later, he applied the principle to all motion, natural and nonnatural (insofar as this distinction still had any meaning), celestial and terrestrial. That second step is as revolutionary as is the first. Galileo, not an especially timid thinker, never took it. But Beeckman did and thereby released himself completely from the limitations of the Aristotelian conception of motion.[15]

THE LAW OF FALLING BODIES

The concept of inertia, like the air pressure theory, proved to be a powerful explanatory principle. Suddenly, all kinds of occult concepts and purely verbal solutions for physical problems appeared superfluous. Beeckman regarded as unnecessary both the heavenly intelligences of Scaliger and the "admirable innate magnetic force" that Gilbert postulated to explain the motion of the Earth.[16]

The planets, including the Earth, simply had no reason *not* to continue in their one-time God-given motion.[17] Intelligences and magnetism were simply superfluous as explaining categories.

A highly significant application of the principle of inertia was the derivation of the law of falling bodies that Beeckman and Descartes worked out together at the end of 1618 (also called the law of free fall).[18] The establishment of a relation between the *distance* traveled by a falling object and the *square of the time* elapsed since the beginning of the fall generally is viewed as "a milestone in the development of modern physics and a major step in superseding medieval ways of thought."[19] The Third Day of Galileo's *Discorsi* (1638) is the locus classicus for this subject, but Descartes's notes from 1618 also have attracted scholars' attention. Therefore, when De Waard rediscovered Beeckman's *Journal* in 1905 and read how an obscure philosopher from Zeeland had helped Descartes arrive at the law of free fall, he immediately believed he should transcribe, edit, and publish the *Journal*. De Waard informed Charles Adam, who was editing the *Oeuvres de Descartes* in Paris and wrote an article on the exchange between Beeckman and Descartes for a Dutch mathematical journal.[20] E. J. Dijksterhuis—who in the decades that followed became one of the most ardent supporters of De Waard's editorial work— focused mainly on Beeckman's contribution to the derivation of the law of free fall. Dijksterhuis wrote several articles and book chapters on Beeckman with a particular focus on his mechanics, as in *The Mechanization of the World Picture*. Alexandre Koyré in France also seemed interested in Beeckman only because of his relation with Descartes and his role as sparring partner of the philosopher who was the first to derive a more or less correct law of free fall. For Koyré, Beeckman's and Descartes's discussion of motion and free fall smashed the closed world of Aristotelian philosophy and replaced it with the open universe of modern science, making it the core of the scientific revolution of the seventeenth century.[21]

Beeckman, on the other hand, never gave the problem of free fall the meaning it would acquire for twentieth-century historians of science. Free fall, for Beeckman, was just one of the many problems his speculations on natural phenomena prompted him to consider. How exactly the fall of a free-falling object took place had intrigued him before he met Descartes. In 1614 he had pondered the question of the path followed by a body shot horizontally—before it hits the ground. Water spurting horizontally from a hole in a water pipe that he had to repair inspired his study of the phenomenon. His attempt to solve the problem by using Christopher Clavius's *Geometrica practica* (1604) proved unsuccessful.

But he took a step toward solving this problem by breaking up the path of the curved stream of water into two components, the horizontal and the vertical motion. The horizontal motion presented no problem: if we disregard the resistance of the medium, this motion is uniform. The vertical component is more difficult: in fact, for Beeckman the problem of free fall was still insoluble. At that moment he could not do much more than formulate (in Dutch) the physical starting point for the determination of the exact behavior of a free-falling object:

> Every thing that falls, falls faster as it falls longer, in the air or in water. So also, if a piece of wood rises in water, it rises faster as it rises longer; this also happens if oil passes through pipes in the water, or smoke through chimneys in the air. Because it preserves its natural motion [tocht] upward or downward to its original extent and in addition it gets another motion [tocht] from the flight [vlucht], on the basic assumption: what moves once, always moves, unless it is impeded, and therefore, as a result, the second flight [vlucht] is larger than the first, and the third faster than the second.[22]

By suggesting that the additional motion is a result of the movement of the object itself, Beeckman still seems to adhere to Nicole Oresme's interpretation of the impetus theory. According to this interpretation, the impetus impressed into a falling object at the outset also assists in producing additional motion, thereby accelerating the fall of the object.[23]

However, in the spring of 1618, when Beeckman inspected several fountains in Brussels, he reformulated this interpretation. He switches to Latin midway through his Dutch text (as if he were transferring from the world of the artisan to the world of learning and wanted to be more precise in his expressions) and observes that a falling stone undergoes acceleration: "The cause is that two motions are combined. First the natural motion downward. Subsequently the stone, having been put into motion, retains this motion and then another natural motion is added to it."[24] Here the source of the additional motion is not found in the moving object, but in an outside cause. Beeckman does not identify this cause, but he understood it to be the unspecified attraction of the Earth, that is, the particles of world ether continually pushing down on the falling object.

Beeckman's meeting with Descartes in November 1618 finally offered him the opportunity to discuss mathematically the motion of falling bodies. Beeckman began his notes by stating the result at which he had arrived in the preceding years:

Objects move downward toward the center of the Earth, the intermediate space being a vacuum, in the following manner. In the first moment [primo momento] so much space [tantum spacium] is traversed as can be by the attraction of the Earth. The object continues in this motion in the second moment and a new motion of attraction is added, so that in the second moment double the space is traversed. In the third moment, the double space is maintained to which is added a third space resulting from the attraction of the Earth, so that in one moment a space triple the first space is traversed.[25]

The argument continues at the bottom of the page in the *Journal* (fig. 4): "Since however these moments are indivisibles [individua], you will have a space through which a body falls in one hour (i.e., such as ADE). The space through which it falls in two hours, doubles the proportion of time (i.e., ADE to ACB), which is the double proportion of AD to AC."[26] In contemporary terminology, this was the way to express a quadratic relation. Linear functions (like $y = ax$) were unknown, and instead mathematicians had learned from Euclid to express the mutual dependence of magnitudes in term of proportions: $y_1 : y_2 = x_1 : x_2$. So the argument quoted above is equivalent to saying that a freely falling body drops a distance that is proportional to the square of the elapsed time.

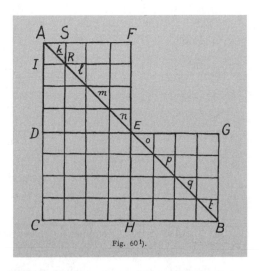

Fig. 60 ¹).

Figure 4. The demonstration of the law of free fall (editor's reconstruction of a very crude drawing by Beeckman, *Journal*, 1:262)

Then Beeckman proceeds with the proof:

> Let the moment of space, through which something falls in one hour, be of some magnitude, say ADEF. In two hours it will go through three such moments, namely AFEGBHCD. But AFED consists of ADE with AFE, and AFEGBHCD consists of ACB with AFE and EGB (i.e., the double of AFE). Thus, if the moment be AIRS, the proportion of space to space is ADE with KLMN to ACB with KLMNOPQT (i.e., double KLMN). But KLMN is much smaller than AFE. Since therefore the proportion of space traversed will consist of the proportion of triangle to triangle, with some equal additions to each term, and since these equal additions become smaller as the moments of space become smaller, it follows that these additions would be of no quantity [nullis quantitatis] when a moment of no quantity is taken. But such is the moment of space through which an object falls. It remains therefore, that the space through which something falls in one hour is related to the space through which it falls in two hours is as the triangle ADE is to the triangle ACB.[27]

Beeckman concludes by saying that this derivation is not his own but comes from Descartes. "This has thus been demonstrated by Mr. Peron, in response to my question whether one could know how much space a falling body would traverse in one hour, if it is known how much it traverses in two hours, taking as given my principles that what is once in motion, will remain in motion, in a vacuum, and supposing that there is a vacuum between the Earth and the falling stone."[28]

Beeckman regarded the derivation as an approximation. He was convinced that it was not correct to suppose, as Descartes did, that the minimum of space traversed has no quantity. Now, if one then proceeded from the physically more correct presupposition that the minimum space traversed has a certain length, an arithmetical series would be produced in which the sum of the numbers contained in half of the terms will never quite relate to the sum of all terms in a ratio of 1:4, even though this relationship will be approximated ever more closely as the number of terms increases. Descartes's solution therefore was, mathematically speaking, a more elegant solution, and Beeckman was prepared to accept it for that reason:

> The proportion of the triangle appeals to us not because there really is no certain mathematically divisible minimum of physical space in which a minimal physical attractive force [minima physica vis attractiva] moves an object (this force is namely not truly continuous but discrete and, as one says in Dutch, "sij trect met

cleyne hurtkens" [she pulls with small jerks] and therefore the foresaid increases consist in a true arithmetical progression). But, I say, it is appealing because this minimum is so small and insensible that because of the multitude of the terms of the progression, the proportion of the numbers does not differ sensibly from the continuous proportion of the triangle.[29]

Whether acceleration is discrete or continuous does not really matter, because the difference in practice will remain below any sensible threshold. Practically and empirically, Beeckman's and Descartes's solutions are equivalent.[30]

Some historians deny Beeckman the credit of having formulated a law of free fall because he accepted the solution only as an approximation and understood that, in reality, things happen a little bit differently. Peter Damerow and Gideon Freudenthal claim that "the quadratic relation celebrated by historians is . . . not Beeckman's law of free falling bodies but rather contradicts it bluntly if taken as a physical theorem."[31] Now it is true that his argument is flawed in more than one respect. Beeckman's "law" of free fall, for example, presupposes a vacuum, while at the same time the space (at least above the falling object) has to be filled with material spirits or particles that push the object toward the Earth.[32] Beeckman did not see this contradiction or did not think it worthwhile to elaborate on it, so we will never know what kind of ad hoc arguments he might have developed to fix the problem.

Beeckman and Descartes each made an indispensable contribution in the derivation of what, in spite of Damerow and Freudenthal's purism, is still widely regarded as the law of free fall. Descartes supplied the mathematical aid, the graphic representation, and the infinitesimal approach. As a result of his studies with the Jesuits at La Flèche, Descartes was better equipped in this area than Beeckman, who was largely self-taught in mathematics. Beeckman had enough mathematical understanding to analyze the problem correctly and to understand Descartes's solution, but he lacked the talent, the proficiency, and imaginative flair of his French friend to arrive at that more elegant formulation on his own. His atomistic natural philosophy, according to which nature is characterized by discontinuities, was too much of an obstacle for him.[33]

On the other hand, we must take into account that Beeckman possessed an understanding of the fundamentals of physics, which, in combination with the mathematical talents of Descartes, proved most fruitful. Above all, these included the principle of inertia and Beeckman's conception of the cause of the motion of falling bodies. Beeckman started from the idea that a stone falls under the influence of a discontinuously working force, which he labeled "the attraction of the Earth" (tractio Terrae) or "attractive force" (vis attractiva).[34] Naturally, this is emphatically not a force of attraction working at a distance. Beeckman's natural philosophy had no place for such a force. Beeckman used the terms only as a common expression for a process whereby the particles of ether, coming from world space and continually striking the Earth and colliding with all earthly objects, are actually responsible for the motion of falling bodies. It would have been more precise if he had spoken not of "attractive force" but rather of "collision force of ether particles." "Weight," he would later write emphatically, "that is to say the downward movement [of objects] toward the center of the Earth, arises from the collision of igniculi, which go from the surrounding atmosphere toward the center of the Earth."[35]

The fact that Beeckman and Descartes based their arguments on physical principles made their approach substantially different from that of Galileo in his *Discorsi* of 1638.[36] Galileo worked exclusively in a kinematic mode: he constructed a mathematical form of the law of falling bodies in order to compare the results with experimental data. Thus, he arrived at a description of how bodies actually fall but did not give an explanation of why they fall this way and not otherwise. By contrast, Beeckman worked dynamically: he based his thinking in part on considerations of the forces that cause the falling motion. Beeckman and Descartes were so sure of their basic principles that they did not test their views against experimental data. Besides, Beeckman realized that experiments would be misleading because one of his basic assumptions, the existence of a vacuum, did not apply in the real world. In any experimental test, air resistance would cause significant deviations. Yet he believed (as he repeatedly told Mersenne years later) that this had no effect on the correctness of his procedures, which entailed first deriving an ideal law on the basis of natural philosophical axioms and then adjusting it as necessary to account for circumstances in the real world.

Beeckman expressed his appreciation for Galileo's work, despite this difference in approach, but the situation was different for Descartes. His position on the derivation of the law of falling bodies was remarkable in any case. Although Descartes knew that the derivation that Beeckman recorded in his *Journal* (and which Descartes gave him) was essentially correct, on more than one occasion Descartes offered fundamentally different interpretations of the law. While Beeckman showed in the *Journal* that there is a relation between the space traversed and the time elapsed from the start of the falling motion, Descartes sometimes assumed that the *velocity* of a free-falling object increases proportionally with the space traversed. This is not correct, but—according to

Descartes—this was equivalent to the other formulation.[37] On the question of whether Descartes's ontology really differed from Beeckman's, and if so, why, historians still differ, but this is a topic that belongs exclusively to the study of Descartes's theory of motion, not Beeckman's.[38]

Still, it is interesting to explore why Descartes later more or less disavowed his contribution to the derivation of the law of falling bodies. With his mathematically and more precisely infinitesimal approach, Descartes helped his Dutch friend overcome his physical scruples, but after 1630 Descartes began to criticize the law for being just a mathematical approximation, an idealization of what actually happened in nature. After he developed an all-inclusive worldview (wherein he explained everything by the motion of small material particles, which fill all of space), Descartes put more emphasis on disturbances caused by the medium, which he had regarded as secondary in 1618. The law of falling bodies, as an idealization of a part of reality, ultimately had a very limited significance for Descartes; actually, in his opinion, it was simply wrong.[39]

MUSIC AND ACOUSTICS

The earlier historians of science—Koyré, for example, and Dijksterhuis—devoted much attention to Beeckman's and Descartes's law of falling bodies. However, as stated before, Beeckman did not consider that law as essential to his philosophy. After 1618, Beeckman never again referred to the law as such (although falling motion is repeatedly mentioned in the *Journal*). Thus, it would be misleading to view this law as the example par excellence of Beeckman's physical and mathematical philosophy. He attached much more value to his contributions to acoustics and, in particular, to his research into the connection between the length of a vibrating string and the pitch of the tone produced.

Acoustics and especially the study of musical harmonies remained a normal and even central component of science well into the seventeenth century. Music had long been a regular part of the quadrivium, equal in significance to arithmetic, geometry, and astronomy. Leading mathematicians and natural philosophers such as Giovanni-Battista Benedetti, Galileo, Kepler, Stevin, Descartes, Christiaan Huygens, and Isaac Newton also carried out research in this area. Their approach to problems of musical theory was mostly in accord with their general approaches to the study of nature, and this tendency applies to Beeckman as well.[40]

Beeckman's most important contribution to acoustics was his proof that the length of a string is inversely proportional to the number of vibrations (the frequency) and thus to pitch. Around 1563, Benedetti already had written (in an unpublished letter) that the number of vibrations doubles when the length of a string is cut in half, and Aristotle had asserted that higher tones are produced in less time than lower tones. But the exact proof of these statements had not yet been established.[41] Beeckman provided the proof in late 1614 (fig. 5). Let the string ab, so his argument begins, be split into two equal parts at point c. Then ab to cb will sound as an octave. And let ab be of such a nature that it can be stretched to h, so that the same string ab can be stretched and made longer to ahb. The half of the string ab will be of such a nature that it can get the same length as the half of ahb and be stretched in the same manner as ab was. Therefore, the string cb will be the same as clb, and clb is the stretched string cb. Then follows the crucial step in the argument:

> Because clb is half of ahb—from the construction it [clb] appears to be equal to hb, which is equal to ah—it follows that hc is double the length of lm: such that bl is to bh, as lm is to hc. Because the nature of the string clb is not changed more or less than that of string ahb, they both strive with the same effort [aequali nixu] to the places of rest ab and cb, and if they pass by these, they return with equal speed. But since hc is double lm, point l passes over the place of rest m twice in the time that h passes the point of rest c once. Since the motion is fastest and most powerful in c and m—the string is at rest in h and l—the movement will be more forceful, the sound will be stronger insofar as the string is farther away from the pause point [locus pausae] (l and h are in any case the pause points between independent tones). The string cb or clb thus produces a tone twice in the same time that the string ab produces only one tone.[42]

Beeckman then continues with the analogous proof for other consonant intervals, such as the third and the fifth. Beeckman also immediately generalized his findings for sounds that were produced not by strings but by other sources, such as the human voice.

The importance of the cited passage should not be underestimated. Certainly, it still contains some striking traditional elements. Thus, Beeckman remains faithful to the Aristotelian idea that—between two opposing motions (of a vibrating string, a reflected beam of light, or an arrow deflected by a wall)—there is always a brief moment in which the moving objects are at rest.[43] Meanwhile, Benedetti and Galileo had already denied the existence of such a point of

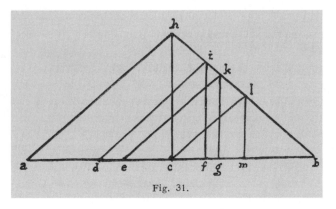

Fig. 31.

Figure 5. Beeckman's proof of consonances (editor's reconstruction of a messy drawing by Beeckman, *Journal*, 1:55)

rest or "quies media."[44] Yet this passage contains the first known mathematical proof of a proposition in acoustics. Not only did Beeckman prove that frequency is inversely proportional to the length of the string, but he also made the connection between the strength (or volume) of the sound and the amplitude of the vibration. Furthermore, he laid the foundation for another important proposition, which states that pitch increases with frequency. It was known that pitch was inversely proportional to the length of the string, and because Beeckman had now proved that string length was inversely proportional to vibration frequency, he had also proved that pitch was proportional to frequency. Beeckman appears to have established this theory in the spring of 1618.

The proportionality of pitch and frequency at that time forms the starting point for his proof of the isochronism of pendulum motion (the phenomenon that, in a given pendulum swing, the movement is always completed in the same time, independent of the amplitude).[45] Beeckman had become interested in pendulum motion during a church service in 1618, when he sat looking at the to and fro motion of lamps.[46] He had noticed that the period of motion was not dependent on the width or "amplitude" (amplitudo), but on the length of the pendulum (later he thought that this movement would go on indefinitely in a vacuum).[47] Beeckman's theoretical treatment of this phenomenon displays many weaknesses, but he was able to think of many applications for a continuously regular pendulum, such as a metronome, a pendulum with a swing of exactly one second, to be used as a universal measure of length, a pulsilogium for the measurement of the human pulse, and a pendulum clock to determine longitude at sea.[48] Again, however, he restricted himself to just thought experiments.

Beeckman's derivation of the inverse proportionality of length and frequency was particularly significant. He realized the fertility of his physical-mathematical philosophy in that simple mathematical proof of the mutual dependence of two physical quantities—one of which, frequency, had barely been conceived before that time. Beeckman situated his critique of Bacon's insufficiently mathematical method and Stevin's insufficiently physical method precisely in the context of musical theory. In May 1628 Beeckman read Bacon's *Sylva sylvarum*, where the English chancellor had recorded his most detailed ideas of musical theory. In response to Bacon's remark that the cause of the octave remained obscure, Beeckman noted in his *Journal*: "I however believe I have found the cause many years ago, as you will see here [i.e., in the notebooks]. For I believed that Bacon was not sufficiently practiced in the combination of mathematics and physics; Simon Stevin, on the other hand, in my opinion was too much addicted [addictus] to mathematics and joined physics too little to it."[49] Beeckman may have been referring here to Stevin's theory of the equal division of the octave as developed in the second volume of his *Wisconstighe Ghedachtenissen* (1608), where he proposed dividing an octave into six whole notes or twelve half notes.[50] Stevin had arranged this system in a purely mathematical way, without taking into account the fact that it was in conflict with musical practice; for Stevin, the mathematical elegance confirmed his theory.[51] In 1629 Beeckman wrote to Mersenne that initially he had been attracted to Stevin's system but gradually realized that the system had no value in musical practice. Practicing musicians used a much simpler system of tonal division, and while this may not have been entirely satisfactory from a theoretical point of view, they at least knew how to make music. Beeckman had begun to see purely geometric constructions such as Stevin's as Pythagorean trifles—"nihil nisi pythagoricum et nugatorium."[52]

Beeckman's ideas on the nature of sound were analogous to those he had on light and magnetism: sound also had a corpuscular nature. The source of sound, a vibrating string for example, divided the surrounding air into small round particles or globules, which were sent off in all directions by the string. Some of these particles reached the human ear and caused the sensation of hearing. Sound thus consisted of air particles in motion.[53] Beeckman was aware of the theory that sound was caused by waves ("pulses" would be a less anachronistic word), but he rejected this theory without giving it serious consideration.[54] If sound was a pulse motion, it would be something like wind, involving not just a few particles but the entire volume of air. It also appeared clear enough that, while wind could come only from one direction, sounds could cross each other in opposing directions.

As a result, the pitch, loudness, and quality of the tones produced could be derived fairly easily from the particle model. The size of the particles determined pitch: larger particles cause lower tones, smaller particles higher ones.[55] For a tone that lies an octave higher on the scale, the air particles are twice as small, a point Beeckman explained by the frequency of vibration. At the higher frequency and thus higher pitch, the vibrating string splits the same quantity of air more often than it does at the lower frequency.[56] For tones that are spaced an octave apart, the vibration is twice as great. The existing, resting air particles are thus divided more finely insofar as the frequency is higher. The loudness of the sound further depends on the quantity of air particles that reach the ear. Beeckman attributed the difference that evidently exists between the sound made by a trumpet and that of a human voice (of the same pitch and loudness) to the different shapes of the globules emitted.

Beeckman also offered an original explanation of the phenomenon of harmony on the basis of his corpuscular acoustic theory. Sound is produced when the vibrating string cuts through the air, but during the movement of the string there are also moments, the *loci pausae*, when the string is briefly at rest, producing silence, not sound. Thus, for every pitch there is a different pattern of sound production and the moments of silence in between. Sounds harmonize with each other, Beeckman explained, whenever their sound-silence patterns coincide. The human spirit experiences as agreeable sounds that coincide in a simple and regular pattern, although how this happens is an entirely different question. Unison is thus the most agreeable sound, followed by the octave, and so forth.

It is not necessary to give a detailed treatment of all possible problems to which this apparently simple explanation of musical harmony might be applied to obtain a general picture of Beeckman's natural philosophy. Other aspects of Beeckman's musical theory require only brief mention. These include resonance, the tuning of musical instruments, the speed of transmission of sound, and psalm correction (i.e., the phenomenon that the people in church sang the psalms better than the crude system of musical annotation suggested).[57] Beeckman regularly recorded his observations on these intrinsically interesting topics, always distinguishing his physical approach from the mathematical approach of Kepler or the experimental approach of Mersenne. Only in Beeckman's notes does musical theory really become the science of acoustics. But apart from their contents, there is another reason to pay attention to Beeckman's extensive treatment of musical theory. When in 1630 Beeckman wanted to make clear to Descartes that he, Beeckman, certainly was capable of independent thought, he pointed specifically to his musical theory. In a letter to Mersenne, Descartes belittled Beeckman, pigeonholing him as a conceited schoolmaster who had only copied from Aristotle. Not only was this a striking misrepresentation of what was actually the case, but Descartes also failed to mention how much he himself had taken from Beeckman in writing his *Compendium musicae* of 1618.

THE INFLUENCE OF THE MEDIUM

Beeckman was aware that his reflections on the motion of material objects were abstract, and applied only to motion in a vacuum. In real life, he understood that this condition could never be fulfilled. Thus, to give the laws of motion some practical value, he adjusted them for situations in which the medium, most often air, had an influence.[58]

Beeckman always emphasized the difference between abstract principles of motion and the actual behavior of moving objects. He had declared that this principle was valid not only for rectilinear motion but also for circular motion. He realized, however, he could not observe the latter in practice. A stone (CF in fig. 6) that is loosely placed on the rim of a turning wheel does not persist in the original circular motion but flies off. In order to explain this phenomenon, Beeckman called upon the force of air resistance. In a thought experiment, he split the stone into two equal parts: one half that faced the middle of the wheel (the inside) and the other half that faced the edge (the outside). With the turning of the wheel, the outermost part has a greater absolute velocity than the innermost part and is thereby in a better position to overcome air resistance than the slower-moving innermost part. The result is that the position of the stone changes: the outermost part comes to lay forward, the innermost part at the back. Jointly, they choose a path that will give them equal velocity—a straight line. Consequently, the stone flies off the wheel.[59] The circular path could have been maintained if all parts of the stone were to carry out that same motion. Beeckman was acquainted with motion of that kind. He thought, for example, of a child's top, which may turn on its own axis as well as around a point on the floor. He also thought of church lamps, which not only rotate as a turntable around their centers but also revolve around the overhead point of attachment in the church ceiling. In these situations, a part that is first on the outside of the circle shortly

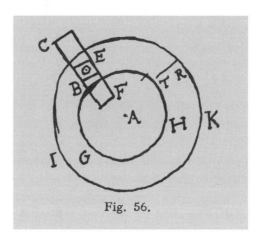

Fig. 56.

Figure 6. Centrifugal phenomena explained by
Beeckman (*Journal*, 1:254)

afterward moves to the inside, and so on. In this manner, he imagined that the
planets carry out their eternal circular motion around the sun: the child's top is a
small-scale mechanical model of the Earth moving around the sun![60]

Air resistance also plays a role in the motion of falling bodies. In his 1618
discussion with Descartes, Beeckman assumed that a vacuum existed between
the falling object and the surface of the Earth, although he knew that the actual
motion of falling bodies in air proceeds differently. Thus, he noted in the spring
of 1618 that, in such a case, acceleration is not unlimited: at first, the velocity
increases until it reaches a point where "the resistance of the air is as great as
the [added] motion; thereafter the rest of the distance is completed at the same
velocity."[61] Beeckman attributed the attainment of a terminal velocity to the fact
that an object falling with increasing velocity continually encounters more air
particles in its path, which must be pushed aside. The increase in resistance at
a certain point neutralizes the increase in velocity; that occurs at the "equilibrium point" (punctum aequalitatis), a concept that appears in his 1617 notes.[62]
Beeckman understood that the answer to the question of where the equilibrium point lies is different for each object because resistance is dependent on
the surface area that collides with the air particles, whereas the quantity of motion depends on the quantity of matter and thus on its volume.[63] The size and
shape of an object are significant in determining the equilibrium point because
the relationship between surface and volume in a large object is different from

that in a small object. A large body, owing to its relatively large volume, will reach its equilibrium point later than a small body.[64] Thus, large objects reach the ground sooner than small objects, because of their higher terminal velocity, although the difference is not as great as the Aristotelians asserted. That difference is sometimes barely observable, because—as Beeckman later pointed out to the skeptical Mersenne—at least in a dense medium, such as water, the equilibrium point is reached shortly after the falling motion begins.[65]

Beeckman not only maintained that there is an equilibrium point; he also thought there is a way to determine that point for a given object. In 1629 he wrote to Mersenne that with the help of a balance he thought he could establish experimentally when that point would be reached.[66] He proceeded from the assumption that as the velocity of a given object increases, its capacity on impact to lift another object on a balance also increases. If one tries to drop the first object from increasing heights, at a certain point one notices that additional height does not result in greater lift capacity: from that point on, the final velocity no longer increases, and the equilibrium point has been reached.[67] Beeckman had not tried such experiments: performing experiments was not his primary methodology. Beeckman's observations of the slower falls of feathers and similar objects through the air—along with prudent reasoning— had convinced him of the existence of an equilibrium point.[68]

LAWS OF COLLISION

A theory of collision is a crucial part of every mechanical philosophy: it is essential to know what happens when objects collide if particles of matter can influence each other only through contact, by pressure and thrust. Beeckman also pondered this phenomenon.[69]

Beeckman proposed seven rules for colliding objects in the midst of intensive discussions with Descartes in December 1618. He divided collisions into two categories: the first involved situations where one of the objects was at rest; the second, where both objects were in motion. Beeckman then reduced cases of the second type to the first type, in order then to establish how collision occurred. He began with rules for the first category. In a vacuum, he stated, a very large body can be moved even by a very small force. This can be seen in the case of a weight that hangs from a long cord, because this weight can be moved even by a "small force" (levi momento). This applies to any body at rest that is struck by a body in motion. Beeckman then continued:

On impact, the resting body will move according to this rule: if both bodies have equal corporality [corporeitas], each of them will start moving twice as slowly as the originally moving bodies. After all, when so many parts are at rest and so many [other] parts are in motion, and when the first receive an equal velocity as the last, that is, when the same impetus [!] must feed a body twice as large as before, the entire body must continue moving twice as slowly. This can indeed be observed in all mechanical tools: a double weight is lifted twice as slowly as a single weight when it is raised by the same force [aequali vi].[70]

From this, it is clear that Beeckman took the principle of the conservation of motion, understood as the product of quantity of matter and velocity (mv), as his point of departure. (The passage also constitutes a remarkable attempt to bring static and collision phenomena under one rubric.) The quantity of motion must be the same before and after the collision. Later, this axiom would play an important role in Descartes's rules for colliding bodies, outlined in his 1644 *Principia Philosophiae*. Yet there are important differences between Beeckman's and Descartes's principles, which the two men probably did not discuss in 1618, when Descartes only listened to Beeckman's ideas. One difference, for example, is that Beeckman understood the quantity of motion as the product of the quantity of matter and the velocity, and not the product of the size of the body and its velocity, as Descartes did (for Descartes, size and mass were interchangeable). Beeckman was deliberately using a masslike concept, which implied the existence of varying densities of matter.

Yet Beeckman was not always consistent in using the concept of "corporeitas," which is not surprising given the nature of the *Journal*. These were not finished and polished discourses, but notes on work in progress. Nonetheless, he certainly knew the difference in principle between volume and mass, as appears in a 1629 reformulation of the previously stated law of colliding bodies. "Whatever moves in a vacuum and meets an object of the same weight (that is to say, of the same quantity of corporality [eiusdem quantitatis corporeae], whether that object takes up more or less space, because this consideration is irrelevant in a vacuum), continues to move together with the previously resting object at half of the original speed."[71] Thus, the concept of "size" in Beeckman's laws of colliding bodies always means "weight" and is always linked to the quantity of matter.[72]

After discussing several other special cases of collision between a body at rest and a body in motion, Beeckman's 1618 text treats the cases in which both bodies are moving. First, he deals with two bodies that are equal and move with equal speed before they collide with each other. According to Beeckman, they immediately come to rest, whereby the motion in each of them is ended. But in case they are moving in the same direction with different speeds, the motions are added together, and they continue to move at half of the total speed. Furthermore, if they are moving in opposing directions and strike each other, the smaller speed is overcome by the larger, and they continue to move at half of the excess rate of motion, in the direction in which the faster body was moving, the reason being that the smaller motion is ended and the remaining portion is divided over both bodies. Then Beeckman continues with the case in which two unequal bodies collide: "If now a body twice as large collides with another body that has the same speed, the greater body loses half of its speed, because it carries the other along with it. The remaining half is divided equally over the two bodies, and each body moves four times as slowly as the larger body was originally moving."[73]

If we disregard the careless use of language (Beeckman here and there uses the word "speed" when he means "the quantity of motion") and an error in the last sentence ("four times as slowly" should be "three times as slowly," that is to say "at one-third the original speed"),[74] it again appears from this passage that Beeckman started from the principle of conservation of the quantity of motion. Furthermore, Beeckman remarks that opposing equal forces cancel each other and that the remnant or excess of unequal motions after the collision is divided over both bodies. Also, when the bodies are in motion in the same direction but of unequal size, the resultant motion is the average of the original speed. These statements suggest that Beeckman took into account the vectorial character of the motion of bodies. The concept of total quantity of motion did not refer to the arithmetical sum of the absolute values of the motions but to the algebraic sum, that is, the sum that takes into account the positive or negative sign of the motion.

Without being able to formulate it, Beeckman was coming to the point of view, later expressed by Huygens, that the common center of gravity of colliding objects before and after the collision continues in the same direction and at the same speed. Huygens stated this proposition in his *De motu corporum ex percussione* (written in 1656 but not published until 1703) as a correction of the rules of collision of Descartes, who had understood the quantity of motion as the product of the size (already an error) and the absolute value of the speed. Huygens showed that in a collision the sum of motion regarded in this manner can remain the same, decrease, or increase, and that a law of conservation of motion is valid only for the algebraic sum of the motions. This correction of

Descartes's theory was a better-supported reformulation of Beeckman's position, from which Descartes had distanced himself after 1618.

According to De Waard, Beeckman's rules of colliding bodies were valid for soft, inelastic bodies such as two pieces of clay. There is no rebound under such circumstances—in contrast with Descartes's laws of collision, which deal with hard objects.[75] This reading of Beeckman's remarks, which implies that they are correct, however, conflicts with Beeckman's own notes. In the middle of a discussion of his rules in 1620, Beeckman states: "I am dealing with bodies that have no pores, or rather in general those which do not rebound, because if that were so [i.e., if they did have pores], they would be in a state to rebound, so that after collision each of them would return with nearly the same speed with which and along the same path on which they encountered each other."[76] Clearly, for Beeckman nonporous bodies were hard, nondeformable objects. As he had already noted in 1613 or 1614, a body could be deformed only if it was porous. "Whatever has no pores," he then wrote, "appears impossible to break; no sharp edge can cut into it if the particles are unable to make room for it."[77] Beeckman thus meant his rules to apply, like those of Descartes, to hard, inelastic bodies, which places the differences between Beeckman and Descartes in even sharper focus.[78] And Beeckman did not choose those bodies arbitrarily. Hard, inelastic bodies in the form of atoms are the fundamental building blocks of the world. If Beeckman had taken soft, elastic bodies as his starting point, his rules of collision would not have produced substantial insights into the nature of the world.

Yet Beeckman remained stuck with a problem. He was aware that his rules, which rested on the principle of conservation of the quantity of motion, with speed seen as a vector quantity, implied that some motion, now stated in scalar terms, could be lost. According to Beeckman, two colliding bodies of equal size moving at equal speed but in opposing directions would, after collision, only come to rest. While at first there had been motion, after collision there was only rest. Beeckman had to accept this consequence on the basis of his rules of colliding bodies, but on other grounds it appeared unacceptable to him. At the end of his 1618 discussions of collision, Beeckman also remarked: "When this proposition has been put forward, one can understand that motion in a vacuum can never result in a faster motion, but that as a result of equal collisions all things tend toward rest. From this it follows that God the Almighty has been able to conserve motion only by moving, at the outset [semel], very large bodies at very high speed. These bodies then continually resuscitate and revive the bodies that tend toward rest." Later he wrote in the margin: "Motion in a vacuum never increases, but always decreases. Why then is there no universal rest?"[79] The gradual disappearance of activity in the world was apparently prevented, according to Beeckman, because at Creation God had given nearly infinite speed to a number of very large bodies.[80] In this manner, universal rest, in principle, could be postponed indefinitely. To what extent Beeckman took "infinite" and "indefinitely" literally is unclear; as a Calvinist, he would not necessarily have objected to the idea of the end of time, that is, the second coming of Christ.

The appeal to God's omnipotent creative power had solved the problem, but as an atomist this did not satisfy Beeckman.[81] Therefore, he tried to think of other solutions, and in 1629 he wondered what would happen if he started from the assumption that the motive force of a body is not dependent on the quantity of matter but only on its speed:

Anyone who takes as a starting point that each body [being at rest] in a vacuum begins to move and continues to move with the motion [i.e., speed] of another, colliding body, would have to explain why all things in the universe do not come to rest. Because if a large body were moved by a very small body at the speed with which the small body was moving (because the latter moved the larger) and if subsequently the larger body, placed in motion, would collide with a small body moving in the opposite direction, then both of them would come to rest; [as if] at first the motions were multiplied and then decreased.[82]

The original decrease in motion would at a later moment be compensated for by an increase in motion, which is what Beeckman proposes. Yet he did not spend much time on this idea before rejecting it; for it implies that a single atom moving at the same speed in the opposite direction could check the motion of a cannonball.

In 1634 Beeckman formulated yet another solution, which he ultimately found more attractive. The problem disappeared when he started not with hard, inelastic, or rigid bodies but with elastic, compressible bodies:

Two bodies that strike each other with equal force [vis] come to rest, unless they rebound; but they come to rest in such a manner that they are compressed together, with the result that on another occasion they would disengage and begin to move with greater speed than this occasion would have caused of itself. So that with the rebound of all parts taken together, the original motion is conserved and renewed, just as a bent steel plate springs back at a given moment. Thus motion once created by God is preserved in eternity, no less than matter [corporeitas] itself.[83]

In the margin he added: "How motion does not have to be destroyed."

These are truly significant remarks. In antiquity, natural philosophers were completely unaware of the incompatibility of atomism and the phenomenon of rebounding. Beeckman was the first atomist to comprehend that positing absolutely hard bodies (atoms) could not be made consistent with the phenomenon of rebounding after collision and the conservation of the quantity of motion. Either atoms are elastic (in that case they are not really atoms), or motion must be lost. At first, Beeckman chose the latter alternative; later he inclined to the former and questioned the axioms of classical atomism. By discarding the idea that his rules of collision would apply to absolutely hard, nonelastic bodies, and by assuming that two bodies traveling at equal speeds that collide head on come to rest only momentarily in the first phase of a collision, which was followed eventually, by a second phase of rebounding, Beeckman thus could confirm what he had written in 1620: "Nothing of motion or of matter is lost in the world."[84] Thus, Beeckman prefigured Isaac Newton, who in his *Opticks* stated the problem in a way that could have been derived directly from Beeckman: "If two equal bodies meet directly in vacuo, they will by the Laws of Motion stop where they meet, and lose all their Motion, and remain at rest, unless they be elastick, and receive new Motion from their spring."[85]

CONCLUSION

As the historian Floris Cohen once remarked, Beeckman was not blessed with awareness of the absurd consequences of some of his ingenious explanations.[86] For instance, his insistence on the corpuscular nature of all natural phenomena prompted him to believe that sound indeed was the result of propagation of a few small particles from the mouth of a speaker to the ears of everyone listening to him. Despite his strong preference for concrete, almost tangible images and models, Beeckman's approach to the problems of physics was in fact often speculative and rationalistic. His "system" was actually a mixed bag of observation, analogy, speculation, and even fantasy.[87] This is perhaps not just a personal matter, an expression of his personality, but also a reflection of the humanist practice of storing knowledge in a commonplace book. A commonplace book simply accumulates quotes and comments from all kinds of sources, treating them as equally trustworthy. Internal tensions and contradictions go unnoticed and critical remarks sit side by side with the most bizarre stories. A commonplace book is thus extremely tolerant of cognitive dissonance. In this sense, there is nothing unusual about Beeckman's ideas. All systems of thought that we label as mechanical philosophy have been saddled with their share of inconsistencies, incoherencies, and contradictions. Even Descartes, who was so proud of his systematic way of thinking, in practice introduced one ad hoc explanation after another as he explained natural phenomena in corpuscular terms. Descartes denied the existence of atoms in his metaphysics, which he needed in his physics.

Now Beeckman, of all mechanical philosophers, may be excused for the unsound and sometimes absurd consequences of his mechanical philosophy because he probably was the first among the moderns to actually develop such a philosophy of nature. Perhaps this is overstating the case. There may have been others working in the same direction. In England, Thomas Harriot, Nicolas Hill, William Warner, and others in the circle of Henry Percy, the ninth Earl of Northumberland, may have preceded him. But their ideas remained unknown to the outside world, so that Beeckman could not have benefited from them.[88] Nor could he profit from the ideas of a host of other contemporary philosophers of nature who were working in the same direction or who had overlapping interests and preferences. Joachim Jungius (born in 1587), Marin Mersenne (1588), Thomas Hobbes (1588), David Gorlaeus (1591), Pierre Gassendi (1592), and René Descartes (1596) were all disenchanted with Aristotelian philosophy and developed corpuscular or even mechanical alternatives, but they all published their ideas after Beeckman had developed his, and sometimes after taking note of Beeckman's ideas.[89] So the point is not that he was the first mechanical philosopher but that he had no examples to follow. Certainly in the first two decades of the seventeenth century, he knew of no other mechanical philosopher. While Descartes in 1618 found a mentor in Beeckman, Beeckman had to find his way by himself.

CHAPTER SIX

Sources for a Mechanical Philosophy

As a mechanical philosopher, Beeckman claimed to be self-taught. He wrote in 1620 that "in philosophy and medicine I have had no teacher whosoever, and in mathematics I had a nonacademic teacher for three months only, thirteen years ago."[1] Earlier, he had admitted to Descartes that he had never spoken with anyone else about his way of integrating physics and mathematics.[2] But Beeckman's mechanistic interpretation of the world must have been based on *some* antecedents. Of course, one can speculate that, as a boy, Beeckman became interested in the behavior of natural objects as he watched his father in his workshop in Middelburg and then simply began to develop his no-nonsense, even materialistic approach to natural phenomena. However, such an explanation seems inadequate: of all the people who were familiar with the arts, why should Isaac Beeckman have developed this rather implausible system of natural philosophy we call the mechanical philosophy? The explanation must entail more than simply watching craftsmen at work.

ANCIENT ATOMISM, MODERN MECHANICS

Beeckman—an avid reader of books, both ancient and modern—most likely had access to the library of Philip van Lansbergen, who must have recognized the talents of his best friend's oldest son during one of his visits to Middelburg.[3] Beeckman does not seem to have read widely in Aristotle—perhaps because Lansbergen was not fond of the ancient philosopher—but he was at home with Galen and a host of other ancient and modern philosophers, physicians, mathematicians, and engineers. Therefore, Beeckman might have gleaned ideas from the books he read during and shortly after his study at Leiden University or during the preparations for the medical degree he earned at Caen in 1618. Beeckman's mechanical philosophy, developed in the years prior to his first meeting with Descartes in 1618, appears to be a bookish philosophy. He had not done experiments or collected a lot of observations, but he had read a substantial number of books about mathematics, natural philosophy, and medicine. Though reading requires decent eyesight, he complained that his poor vision made him unfit for empirical research and medical practice. Beeckman did not recognize the value of good empirical and experimental research until the 1620s and 1630s. Yet, as he read, he constantly commented on the books' contents. One can therefore suppose that Beeckman derived from books the arguments, the concepts, and even some of the experiments he needed to develop his mechanical philosophy and overturn Aristotelian natural philosophy. Books, according to this line of reasoning, constitute a natural starting point to explore the sources of Beeckman's mechanical philosophy of nature.

Two authors emerge as primary sources of inspiration. The first is the Roman poet and philosopher Titus Lucretius Carus (first century B.C.), the famous (or infamous) atomist. Beeckman, with his strictly mechanical philosophy of nature, comes closer to Lucretius and the other ancient atomists than any of the seventeenth-century mechanical philosophers. From a practical standpoint, all proponents of the mechanical philosophy came to realize that a system of natural philosophy based solely on the notions of passive matter and inertia simply did not work.[4] Long before Newton, they therefore introduced or reintroduced active principles and inherent forces that acted without direct contact and thus could operate at a distance—unlike Lucretius and Beeckman. The sixteenth-century Dutch or actually Flemish engineer Simon Stevin, who wrote quite a few books on mechanics and left behind some manuscripts that Beeckman studied carefully, also greatly influenced Beeckman. Mechanics had become a well-respected branch of mathematics in the sixteenth century, and Stevin's work was the culmination of this movement. The fact that Stevin had been an immigrant from the southern provinces, just like Beeckman's father, also may have created a special bond between Beeckman and Stevin.[5] Thus, Lucretius's atomism and Stevin's mechanics may have played key roles in the development of Beeckman's mechanical philosophy.

Anyone interested in philosophy in the early seventeenth century could get a good idea of ancient atomism's contributions. Scholars had various texts at their disposal containing the atomists' theories. Both Aristotle and Galen had argued with atomists and thus made the atomist ideas accessible to later generations. Furthermore, Lucretius's long didactic poem *De rerum natura libri VI* (Six Books on the Nature of Things) had been rediscovered in 1417 and printed in 1473, with many new editions in subsequent years. Lucretius expounds on Epicurus's atomistic natural philosophy in *De rerum natura* with the aim of eradicating superstitious fears of divine intervention in the world and of the punishment of the soul in the afterlife. He does so by demonstrating first that the world is governed by the laws of nature, or—to be more precise—by "pacts of nature" (foedera naturae). Second, he argues that the soul is mortal and perishes with the body. The bulk of the poem is devoted to setting out a detailed atomic view of the universe, which Epicurus had adopted (with modifications) from two other Greek philosophers, Leucippus and Democritus. The atomists' fundamental idea is that of an infinite number of atoms moving through infinite space, and bringing about all things and events in nature by their combinations. This philosophy is purely material, but—in the case of Lucretius—not deterministic. Lucretius postulates free will for man with a certain spontaneity of movement in the atoms corresponding to human free will. The poem also touches upon Epicurus's moral theory—that pleasure is the end of life. Lucretius's visual language—adorning what otherwise would have been a dry exposition of a philosophical system with a wealth of illustrations and vivid imagery—made him popular among philosophers in both ancient Rome and early modern Europe. Beeckman was only one of numerous early modern philosophers charmed by Lucretius's exposition of ancient atomism, Michel de Montaigne being another. As a consequence, the emergence of the mechanical philosophy has often been associated with the revival of ancient atomism. "Mechanical philosophy," William Shea once remarked, "had its roots in the atomism of the ancient world."[6] Beeckman's work seems to offer a perfect illustration of this thesis.[7]

There are at least ten or eleven direct references to Lucretius in the *Journal* and quite a few indirect ones.[8] At times, Beeckman's daily observations stem directly from *De rerum natura*. For example, when Beeckman describes the behavior of the atoms of light by comparing them to dust particles dancing in sunlight streaming through the window, he is quoting Lucretius—a reminder that ancient literature served as a source of observations as well as theories and arguments in the sixteenth and early seventeenth centuries.[9] Beeckman also derived key concepts directly and without acknowledgment from Lucretius. The concept of the *primordia*, so essential to Beeckman's explanation of the effects of medicines and the generation of living beings out of passive matter, represents one example among many others. Beeckman's borrowing words and observations from Lucretius is not surprising. Beeckman was raised in the commonplace tradition and never abandoned humanist habits of mind. Four editions of *De rerum natura* are listed in Beeckman's auction catalog, including the 1595 Leiden edition with comments by Obertus Giphanius.

Beeckman's first reference to Lucretius appears when he worked as a candlemaker in Zierikzee. Until 1615, he had been mainly interested in mathematics and related disciplines such as mechanics, astronomy, and optics, but he then began to express interest in natural philosophy—at the same time that he considered leaving his business to study medicine. In the second half of 1614, he offers a new and purely mechanical explanation of magnetism, postulating corporeal "spirits" that flow through the magnet and cause decreased air pressure between the magnet and the object that is attracted. This is very much like Lucretius's explanation of the same phenomenon, with which he basically agrees. He even supposes, like Lucretius, that the corporal spirits are held together by little "hooks."[10] But does this mean that Beeckman derived his explanation from Lucretius? There are some differences between their explanations. Beeckman invokes the concept of air pressure where Lucretius had written about the wind that pushes the object toward the magnet. But more important is the fact that Beeckman had already formulated his concept of inertia, a major pillar of his mechanical philosophy, around 1610 and had already accepted the existence of the vacuum shortly after, as one can see from his correspondence with Jeremias van Laren. It seems as though Beeckman had arrived at the basic notions of his natural philosophy *before* his study of Lucretius. Ancient atomism only confirmed the conclusions he had already arrived at independently of the ancient atomists. Perhaps he became interested in Lucretius because he was looking for confirmation of his own ideas.[11]

It is also worth noticing that Beeckman distanced himself from Lucretius in several ways. In 1628, for example, Beeckman refuted Lucretius's theory of vision. Lucretius had theorized that "seeing" an object means receiving some sort of print emanating from the object itself, like some sort of "skin," but Beeckman disagreed.[12] He gave the note devoted to this topic the marginal heading "Lucretius refutatus," and as such it was published in the posthumously edited *Centuria*.[13] In general, there is enough evidence to conclude that Beeckman's attitude toward Lucretius was selective, and always had been. Beeckman knew of the similarities and the differences between his own ideas and those of Lucretius, but he never gives the impression that he derived his ideas from the Roman poet. At least in Beeckman's case, the roots of his mechanical philosophy are *not* to be found in ancient atomism.

Simon Stevin's contribution to the development of Beeckman's natural philosophy was more substantive. Of course, there are important differences between Stevin and Beeckman. Stevin never expressed an interest in natural philosophy. He had written extensively about bookkeeping, mathematics, music, mathematical astronomy, the construction of mills and sluices, the art of fortification, hydrostatics and other parts of mechanics, but never about the structure of matter or the explanation of magnetism.[14] Stevin was more interested in applying his knowledge than in the theoretical aspects of the disciplines involved. In his own words, Stevin preferred the "daet" (practice) over the "spiegheling" (theory), although he acknowledged the value of theory as such.[15] Beeckman, on the other hand, definitely preferred to delve into theory. Still, he admired Stevin. He never met him personally, but after Stevin's death in 1620, Beeckman traveled to Stevin's widow's home in Hazerswoude and copied excerpts from Stevin's unpublished treatises on architecture and music.[16] Stevin's work repeatedly served as the point of departure for Beeckman's own speculations. Indeed, on the first page of the *Journal* Beeckman acknowledges that Stevin's writings on astronomy provided convincing arguments in favor of Copernicus's theory. And Beeckman and Descartes's discussions at Breda in 1618 included Stevin's work on hydrostatic problems.

Stevin's writings inspired Beeckman on a more subtle level, too. In 1623 Beeckman discusses the pressure exerted by water in a vessel, a topic he had analyzed before, at times referring to Stevin's work.[17] Sometimes Beeckman added illustrations directly derived from Stevin, picturing water as an amorphous mass represented by little dots or short lines. But now, in 1623, he changed the form of representation: in the illustration accompanying Beeckman's note, only the water directly adjacent to the vessel is represented by a chain of little round particles, the "globuli" (see fig. 7), instead of an amorphous mass represented by little dots.

The similarity with the small circles representing the light particles in Beeckman's somewhat earlier illustration of a ray of light refracting on entering water is remarkable (see fig. 3). But his discussion of the illustration suggests something more. First, Beeckman imagines that the water particles are connected to each other ("si inter se forent connexi"), as if they constituted a real chain. On reflection, however, he realizes that it does not matter whether these particles are tied to each other. The effect of their weight and the pressure they exert would be the same if they simply lay beside each other.[18]

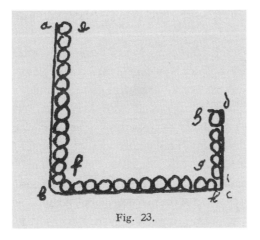

Fig. 23.

Figure 7. Pressure of water in a vessel
(*Journal,* 2:236)

Anyone familiar with the history of sixteenth- and seventeenth-century mechanics will immediately see the parallel with this line of reasoning and the images in Stevin's well-known book, *The Art of Weighing,* published in 1586. In this book, Stevin deals with oblique forces and gives his own proof of the law of the inclined plane—the proposition that two spheres on inclined planes, connected by a string passing over a pulley at the top, will be in equilibrium if the ratio of their weights is equal to the lengths of the planes.[19] Stevin discusses what will happen to a chain of connected spheres hanging from a triangle with unequal sides. He concludes that this chain will be at rest, although there are four spheres to the left and only one to the right of the upper angle of the triangle. Stevin was so proud of this proof of the law of inclined planes that he used the figure illustrating the proof as a vignette for the great majority of his works, including *The Art of Weighing* itself. Stevin's famous dictum "Wonder is no wonder" accompanies the figure in *The Art of Weighing.* Stevin also used the figure as a seal to his letters and as a mark on his mechanical inventions. This "wreath of spheres" was one of the most popular illustrations in early modern mechanics.

By picturing first light and then water as a chain of little round particles, Beeckman gave a natural philosophical interpretation to this iconic figure in the science of mechanics. In doing so, Beeckman crossed the line that separates mechanics from natural philosophy. Whereas Stevin referred to little balls that could easily be seen with the naked eye—in case someone actually performed the experiment with the wreath of spheres—Beeckman's globuli were purely mental constructions; these globuli, although they really existed, were too tiny to be seen and one had simply to assume that they behaved more or less the way Beeckman had described. Beeckman was an atomist, whereas Stevin probably was not, but the resemblance is evident in their way of picturing reality. Stevin certainly did not provide Beeckman with arguments to adopt an atomistic theory of matter, but Stevin's illustrations clearly influenced Beeckman.

What prompted Beeckman to cross this line between mechanics and natural philosophy? Illustrations and pictures do not speak for themselves. They have no fixed meaning but can acquire new meanings in different contexts. No inner logic compelled Beeckman to take Stevin's figure of the wreath of spheres as a representation of the invisible globuli comprising the substance of water and light. New ideas are never simply the sum of all the ideas that inspire them, as we sometimes—carelessly—assume. The gulf separating mechanics and natural philosophy was narrowing throughout the sixteenth century. The 1517 rediscovery of the *Quaestiones mechanicae*—generally taken for a genuine work of Aristotle himself—raised the social and intellectual status of mechanics and its practitioners.[20] God himself is presented as a practitioner of mechanics, an engineer who devised the world according to the rules of mechanics in a 1599 edition by the French mathematician Henri de Monantheuil. This appears to obliterate the crucial difference between mechanics, an art dealing with unnatural movements, and natural philosophy, a part of philosophy dealing with natural movements.[21] But such a conclusion was only implicit in the commentary of Monantheuil and others; to draw this conclusion explicitly required something more. So, although authors like Lucretius and Stevin—and many others—provided Beeckman with essential ingredients for his mechanical philosophy, the fact that this philosophy was an original contribution to the understanding of nature compels us to look for other factors—outside the domain of mechanics and natural philosophy—that led him to adopt a corpuscular view of nature and cross the boundary between mechanics and natural philosophy.

THE CRAFTSMAN'S BACKGROUND

Perhaps Beeckman's craft background is a more obvious source of his mechanical philosophy.[22] Beeckman made candles and installed and repaired water supply systems in the years between 1612 and 1618 when he developed his natural philosophy. Thus, he gained wide experience with practical mechanics and was often confronted with natural phenomena that compelled him to seek

more far-reaching explanations. A craftsman's mentality must have deeply influenced his ideas on natural philosophy.

In reading Beeckman's *Journal*, one immediately notices how the most abstract natural philosophical questions alternate with discussions of highly
practical matters such as the manufacture of better candles or the dredging of
harbors. Beeckman's mind seems constantly to be working on two tracks, as he
shifts between artisan and natural philosopher. Sometimes these two tracks
intersect: for example, when Beeckman shifts from Latin to Dutch or the other
way around, or when he poses a question like a craftsman, not like a mathematician (recall how he phrased his question about the behavior of a free-falling object by asking for specific values instead of a general relation).[23] The artisan also
meets the scholar when Beeckman explains natural phenomena using craft-
based analogies. For example, Beeckman solves a problem raised by Galen regarding how the bloodstream can simultaneously supply nutrients and carry
away wastes when he points out that lighter liquids can rise while heavier ones
fall to the bottom in an ordinary water pipe.[24] Also, Beeckman regularly compares natural phenomena to ordinary objects in his workshop. For example, he
compared the human body and the sun to a "candlewick" (ellichnium): all three
objects consume something (by working or by burning) without being consumed.[25] Even Beeckman's secret admiration of Cornelis Drebbel points to the
craftsman and "mechanick" inside Beeckman.

Most importantly, Beeckman's artisanal background surfaces when he discusses criteria for intelligible explanations of natural phenomena. He regularly
expresses resentment toward abstract and improbable explanations that traditional natural philosophers had offered for certain phenomena. He objected to
concepts such as *impetus* and *potentia rarescendi et densandi* (the capacity to become less or more dense) because he simply could not picture a mechanism that
might explain how natural phenomena were generated, and not because they
were untrue or untenable.[26] For example, in a letter to Mersenne, Beeckman rejected as fables philosophical concepts regarding the necessary unity of things,
the transmission through air of immaterial prints of a seen object, and the im-
possibility of motion in a vacuum. He did not want to admit anything in philosophy unless it could be presented visually to the imagination.[27] Beeckman
believed the world consisted of tangible, concrete things that act upon each other
in a way he could visualize. He views the world like a craftsman inspecting a
machine he is about to repair. No craftsman would explain the functioning of a
simple mechanical device by invoking occult virtues or immaterial causes.
Instead, in the words of John Schuster, explanations in the mechanical arts

normally rest "on the appeal to a clear picture of the structure and interactions of the constitutive parts of the apparatus."[28] Throughout the *Journal*, Beeckman applies the criteria used in the communication between craftsmen and artisans to judge new ideas and new inventions, always appealing, as again Schuster puts it, "to a picturable or imaginable structure of parts whose motions are controlled within a theory of mechanics." This is what Schuster then characterizes as Beeckman's "naive translation of the imperatives of the mechanical arts into the terms of natural philosophy."[29]

Craftsmen and engineers had begun to enter and enrich the world of scholarship in the fifteenth and sixteenth centuries.[30] The development of commerce and industry in late medieval and early modern Europe stimulated new intellectual activities in which both scholars and highly trained mathematical practitioners participated. In Italy, Germany, and the Netherlands, architects, military engineers, and some skilled painters attained recognition as experts in natural inquiry. Leon Battista Alberti and Simon Stevin are two examples of this phenomenon. And men with university training—such as Paracelsus and Galileo—recognized that they could learn a lot from engineers, apothecaries, and even ordinary craftsmen: they descended from their ivory towers and sought out the local apothecary, the blacksmith, or the shipbuilder to learn how these people interacted with nature. The integration of skilled craft and academic knowledge, the hand and the mind, resulted in the empirical and experimental methodology that formed the core of the new science of the seventeenth century. Recent research has revealed the close connection between the working philosophy of the mathematical practitioners, sometimes inscribed in their instruments, and the new understanding of nature as articulated by philosophers in the seventeenth century.[31] Early modern science emerged from the ways of knowing of artisans, painters, chemists, and other practitioners of mechanical arts. This rapprochement between craftsmen and scholars was especially pronounced in the Dutch Republic, where social barriers were not as stark as in more aristocratic societies. Beeckman's mechanical philosophy grew out of his experience as a craftsman: he was both a craftsman *and* a scholar.

But it would be misleading to portray this craftsman-scholar relationship as simple or straightforward. Beeckman was a craftsman, but at times he doubted he could learn anything from other craftsmen. He respected their skills, and—after he became a teacher—never felt superior to them. Toward end of his life, Beeckman worked hard to master the skill of grinding lenses. Yet he had little confidence in craftsmen's schemes to improve existing technology. He received and reviewed many proposals to improve specific technologies—from methods of deepening a harbor to new mechanisms for a horse-driven mill—but in nearly every case the theoretical foundations proved inadequate and the so-called invention was flawed. Certainly, Beeckman knew immediately that the invention was unsound when an inventor declared that his invention rested on the principle of perpetual motion. Repeatedly, he understood that a crafts-man's understanding of the fundamental tenets of mechanics was inversely proportional to the craftsman's enthusiasm for his new "invention."[32] For ex-ample, in 1633 he observed: "The practitioners [practisyns] say every time that they have found an invention by which they can do much more, that is twice, three times, etc., even twenty times more than anyone was able to do before, just as Willebrord Snellius told me that someone at Amsterdam had certainly found out (and Snellius believed that). But they all deceive themselves in the proportionality between time and force."[33] Beeckman, who was trained as a craftsman, was thus more careful in his expectations than Snellius, the current professor of mathematics at Leiden, whose training was purely academic.

Beeckman noticed, however, that when craftsmen received help from some-one schooled in mathematics and mechanics, their work markedly improved. In 1623 a distant relative, who was a silversmith in Middelburg, asked Beeckman to help him enlarge and reduce certain models.[34] The silversmith himself, Vin-cent Everdeys, did not know how to proceed. "Each one goes to his neighbor to borrow a measure, which was already found by trial and error and nevertheless still lacks countless quantities," Beeckman observed. Beeckman used his knowl-edge of isoperimetric figures to help Everdeys.[35] Thus, Beeckman considered starting a Collegium Physico-mathematicum, first in Rotterdam and later in Dordrecht, to teach the fundamentals of Euclidean geometry to "carpenters, masons, skippers and other citizens" and to "all those who were known to be somewhat practiced in mechanical works and other skills."[36] While Beeckman had a realistic understanding of the limits of craftsmen's knowledge, he be-lieved that theoretical education might raise the crafts to a higher level. What we have here is a scholar helping out a craftsman, not the other way around. The somewhat paradoxical example of Beeckman—a craftsman clearly aware of the limitations of craft knowledge—seems to confirm the apparently discarded opinion of Rupert Hall that, insofar as the mechanical and mathematical sci-ences of early modern times profited from the experiences of the craftsman, this was due to the changed attitude of the philosophers and not to any contri-butions by the craftsmen themselves. This conclusion may be too rash, given the many instances in which Beeckman elaborated on his own experience as a craftsman; clearly Beeckman's assessment of craft knowledge and its relevance for the study of nature was astute. Yet Beeckman's reservations regarding the relevance of craft experience throws some doubt on the hypothesis that it was his craft background that set him on the path of a mechanical philosophy of nature. Other considerations reinforce this doubt. In the sixteenth and early seventeenth centuries, mechanical concepts also played a role in magical or neo-Platonic philosophies—that is in essentially nonmechanistic philosophies. Thus, a direct translation of technical experience into mechanistic philosophy seems even more improbable. If, as Schuster argued, Beeckman indeed translated the imperatives of the mechanical arts into the terms of natural philosophy, this translation did not happen automatically. What made Beeckman believe that his craftsman's inspiration or technical intuition could legitimately be turned into a new and distinctively mechanistic philosophy of nature?

THE ROLE OF RELIGION

Isaac Beeckman was a deeply religious man. He grew up in an environment where religious purity counted more than social success: he underwent theological training and served as an active elder in the Reformed Church. The *Journal* also contains several reflections on his religious beliefs. At times, Beeckman expressed doubts about the firmness of his beliefs, which was not uncommon among orthodox Calvinists. Reflecting on his fear of being buried alive, he wrote in 1631: "My unbelief is so immense that I fear that I am by far the weakest believer of all Christians."[37] Yet such moments of doubt strengthened his belief in God, for to him this fear, which he also recognized in his children, was in the end only the manifestation of an instinct that God had implanted in nature as a proof of eternal life after death.[38] Beeckman's command of theology was also impressive and confirmed by the auction catalog of his library. Theological books made up more than a quarter of the books listed in the catalog. This section must also have been the most expensive, given the relatively high number of folios. The book titles testify to his wide reading in theology. Although the majority of texts had been written by Reformed theologians, Beeckman's library included works by the church fathers, Erasmus, Luther, Melanchthon, and even Catholic authors. Yet Beeckman owned twelve books by Calvin, more than any other author.[39] Perhaps Beeckman's Calvinist convictions influenced the development of his natural philosophy.

Beeckman's situation, as a strict orthodox Protestant actively engaged in natural philosophy, is consistent with the well-known but also heavily criticized thesis regarding the relationship between the Reformation and the rise of modern science in early modern Europe. Authors like Max Weber, Robert Merton, and Reyer Hooykaas have tried to establish such a connection. Weber linked the spirit of Protestantism and the so-called disenchantment of the world, especially the purging of nature of magical forces.[40] Merton particularized this very general outline to the thesis that in early modern England the stricter Puritans had provided essential support for the new science.[41] Hooykaas went one step further by establishing a relationship between the life and teachings of the stricter Protestants on the one hand and their rejection of religious (and by implication also philosophical) authority. He also linked Puritanism to a preference for the experimental method and to the emphasis on the contingent character of nature that he thought was typical of modern natural science.[42] Hooykaas viewed Beeckman, a mechanical philosopher who was a member of the Dutch Reformed Church and who sympathized with English Puritans and dissidents, as a prime example of this connection between Puritanism and science.

Some observations in the *Journal* suggest that Beeckman's religious convictions may have played a role in the choices he made as a natural philosopher. For example, in 1626, when Beeckman learned from the Rotterdam regent Van Berckel that burgomaster Puyck might fund a venture involving a new type of horse-driven mill, he immediately warned Van Berckel against the enterprise. Van Berckel had told Beeckman that the invention rested on the construction of a perpetual motion machine, and that, according to Beeckman, was an illusion. "Only God makes living wheels or perpetual motion," he said.[43] Even before he had investigated the machinery, Beeckman knew it would not work on religious grounds. Yet such remarks are vague, and it is hard to establish the role religion and theology played in Beeckman's natural philosophy. Did his Reformed Protestant convictions, for example, *motivate* him to spend so much of his time on natural inquiry, or did these convictions also play a role in his specific natural philosophical *arguments*? Beeckman's ideas about God and his relation to the world convinced him that the horse-driven mill construction had to be unsound, but Beeckman did not use religious notions in his discussions with the inventors. He referred instead to their misunderstanding of mechanics.

What were Beeckman's ideas about man's capacity to understand nature? Calvin had had a rather pessimistic view of the consequences of the Fall of Man, which, according to him, had led to a serious reduction in man's ability to know nature. Yet this in the end only stimulated his followers to try and repair this corruption of human understanding by, for instance, a program of intensive and collective experimental research.[44] Francis Bacon, the English Puritans of the 1640s and 1650s, and finally the founders of the Royal Society based their efforts exactly on such an interpretation of the biblical Fall. But what about the Dutch Calvinists, and what about Beeckman? He had an intimate knowledge of Calvin's *Institutes* and had read at least some of Bacon's major works. And yet Adam and Eve and their fall from grace are not mentioned in the *Journal*. Is this simply because Beeckman did not feel the need to justify his doings by invoking the Calvinist theory about the Fall in his private notebooks, or does this silence reflect the indifference of a practicing natural philosopher to theological theory?

Beeckman never questioned God's role as creator and maintainer of the world. God is continually present in nature, and nature cannot be conceived without God. "God is author of nature itself. And therefore we must ascribe to him all good and weighty matters, whether people can prophesy them or not, whether people do them themselves or not."[45] Indeed, Beeckman agreed with the views of the well-known English Puritan preacher William Perkins, who was very influential in the Dutch Republic and especially in the circles frequented by Beeckman. In *An Exposition of the Creede* (1595), for example, Perkins expressly stated that God is continually working in things natural,

> first by *sustaining* and preserving them that they decay not; secondly by *mooving*
> them that they may attaine to the particular endes for which they were severally
> ordained. For the qualities and vertues which were placed in the Sunne, Moone,
> starres, trees, plants, seedes &c. would lie dead in them and be unprofitable, un-
> less they were not onely preserved, but also stirred up and quickned by the power
> of God as oft as he imploies them to any use.[46]

Although Perkins was an old-school philosopher in the sense that he still believed in the hidden virtues in natural objects, Beeckman would have fully agreed with his idea that God was the carrier and mover of the world—at least on a metaphysical level.[47]

This idea did not preclude the possibility that on a natural philosophical level all phenomena had their natural causes and explanations. To put it even more strongly, the Protestant and certainly the Dutch Reformed or Puritan conceptions of God as the absolute Other, of the absolute gulf separating God and mankind, and of the distance between the active God and the passive world created the very space for explaining nature in its own terms and deriving the phenomena of nature from natural causes. In the last resort, God was indeed responsible for the events in nature, but he allowed these events to take place by natural means, as Beeckman noted in the passage immediately preceding the

"God is the author of nature" quotation above: "Then, nevertheless, great and weighty matters are often rightly ascribed to Him, although they take place by means of a cause that we knew would certainly occur, just as man may well thank God, that He allowed the night to come at that particular hour, by which we received the victory."[48] The fact that a once-mysterious event such as an eclipse can now be explained and even predicted does not mean that we no longer attribute its occurrence to God's will. Beeckman might have found this view of God as the distant cause of everything that happens in the world difficult to maintain had he become a preacher. A preacher has to give comfort to those who are suffering and to lift up others who have lost faith, which he would not be able to do if he did not posit a more direct presence of God in the world. As a natural philosopher, however, Beeckman found the concept of a distant God who always works by natural means very attractive.

Beeckman thus distinguished between God as creator and upholder of the world, on the one hand, and the series of natural causes that can explain the actual phenomena of nature, on the other. Beeckman understood that such secondary causes existed, and that the course of natural phenomena did not arise by chance. As I noted before, Beeckman distanced himself from the theory expounded by the ancient atomists that all things in nature, dead objects as well as living beings, came into being as a result of a "haphazard concourse" (fortuitus concursus). Beeckman argues: "We who say that God made the atoms in an artificial way [artificialiter], believe that their coming together is not arbitrary."[49] In the beginning of time, all things had been "well ordered" (legitime dispositis), and from then on atoms coalesced to form more complex structures according to rules also laid down by God. These rules placed limits on what could happen in nature. In response to a remark by Galen in *De usu partium corporis humani*, Beeckman maintained that in the beginning God had created the "atoms" (minima corpuscula) out of nothing, but had formed them in such a way that not every arbitrary thing could come out of their combination but only the things that had been fixed in advance. "Therefore it is enough to say that all things have arisen according to their nature and the constitution of their place, after God in the beginning had created their principia, which, linked to each other, *could not but produce them* [non possint non hoc facere]."[50] Emphasis is added here because similar phrases appear throughout the *Journal*. They indicate Beeckman's belief in total passivity in nature; nature is not endowed with any creative power but exists only because God created the world and the rules according to which the machine of the world must function.[51] There is even a parallel between the theory that atoms and primordial matter are predestined to come together as a camel or a human being and the doctrine of predestination that is so characteristic of the Calvinist creed to which Beeckman subscribed.

Beeckman thus agreed with Galen, who had refuted the view that God could make all things "regardless of the nature of the material involved" (absque respectu materiae). According to Galen, God could not make a human being out of a stone but only put together material that was already suited for a particular purpose. But Beeckman disagreed with Galen's assertion that the laws by which natural phenomena occur had placed limits on God at the creation. According to Beeckman, God—out of his own free will—had imposed these laws on nature, which theoretically allowed for the possibility that God could also change these laws or temporarily suspend them.[52] As a good Protestant, however, Beeckman believed that, since the coming of Christ, such miracles were no longer necessary and thus no longer occurred.

Beeckman's point of view gave him the opportunity to make a clear methodological distinction between theology and natural philosophy. He warned against prematurely bringing theological arguments into natural science inquiries. When Jeremias van Laren invoked God in 1613 in an intellectual exchange on the existence of the vacuum, Beeckman immediately corrected him. It is not right, he wrote, to say that God freely gave air a capacity for rarefaction and compression, because if that were the case, it would be incomprehensible to the human mind.[53] God made nature completely comprehensible, so it is unnecessary to take refuge in supernatural or theological explanations.[54] And just as theology must remain outside natural philosophical discussions, natural philosophy must stand aside from theological questions. In 1630 Beeckman wrote to Father Mersenne: "Many persons have written many probable things about the world system, infinite space, eternity, the inhabitants of the stars, the empty space between the stars, and sunspots. But who has ever reflected without foolishness on the Holy Trinity and the reconciliation of human freedom with divine predestination? In these matters, the philosophy granted to us comes directly in conflict with theology."[55] Theology and natural philosophy not only had their own domains but were also clearly distinct from each other methodologically. Referring to the motto of Simon Stevin, "Wonder is no wonder," Beeckman wrote in 1626: "In philosophy, one must always proceed from wonder to no wonder, that is, one should continue one's investigation until that which we thought strange no longer seems strange to us; but in theology, one must proceed from no wonder to wonder, that is, one must study the Scriptures until that which does not seem strange to us, does seem strange, and that all is wonderful."[56]

Thus, in natural philosophy there is actually no room for wonder, and if we meet or experience something wonderful, that only means that we do not yet know the true cause of the phenomenon. It is just like seeing a magician, whose tricks we do not understand and therefore regard as "magic," because we do not yet comprehend what he is doing; or like people who are afraid of noises at night, noises that would not frighten them during the day because they then could see what is happening.[57]

> Because this is common among the people: if their experience is contrary to their reasoning, they take refuge in that which cannot be experienced. Thus, they seek the extraordinary in illnesses, as if these had come and were continuing by magic, etc. Thus people also speak of rain, snow, lightning, thunder, etc., of which the causes are hidden for us by the particularities of the circumstances. . . . There is no reason for us to seek any miracle in rain, snow, etc., any more than in the path of the sun or the moon, whose course is known to us.[58]

This attitude toward the miraculous also made Beeckman immune to explanations that called upon supernatural powers, whether from the devil or of a magical nature. He certainly believed in the existence of the devil, "a spirit who occupies no place, but who can pass through wood and stone."[59] Yet the devil did not possess the supernatural powers that were ascribed to him; for example, devils cannot read minds, and, just like human beings, they must proceed on the basis of facial expressions.[60] For Beeckman, attributing certain phenomena to the devil was a sign that one had not yet discovered the true cause.

Beeckman also did not believe in other sorts of magic—from sorcery and magical healing to Cabbala and predictive astrology.[61] Thus, even though he admired Drebbel's inventiveness and tried to figure out how his perpetual motion machines really worked, he was immune to the extravagant philosophical claims with which Drebbel presented his inventions. Beeckman simply attributed the working of the Drebbel instrument—a glass sphere filled with water that supposedly rose and fell in sympathy with the tides—to the common cause of both phenomena, that is, the periodically stronger and weaker radiation of fire particles from the cosmos.[62] In the same vein, Beeckman gave a remarkably detached account of the Gorkum miller Balthasar van der Veen's fanciful ideas, including his claim that he could cure people through knowledge of the Cabbala. Beeckman assumed the attitude of the interested outsider who was able to find such silly ideas (*nugae* or nonsense as Beeckman labeled them) curious because he was long past the stage at which he could have believed in them.[63] While many people in Holland still believed in the existence of sorcery, Beeckman is the prototype of the levelheaded town dweller in whose concrete world enchantment was absent.

Beeckman maintained a similar attitude toward many scholars whose works contained magical, hermetic, or neo-Platonic elements. He had so little confidence in such philosophical systems that he did not even bother to refute them. Beeckman engaged with the ideas of authors such as Gilbert, Kepler, or Bacon only because some of their ideas mirrored his own. But when Gilbert argued that the world moves by means of a wonderful intrinsic magnetic force, Beeckman thought it was the same as ascribing intelligence to the Earth, a view he regarded as "unworthy of a philosopher" (philosopho indignum).[64] When Kepler posited a "shaping nature in the universe" (conformatrix natura in universo), Beeckman called it ridiculous and again unworthy of a philosopher.[65] In the case of Bacon, Beeckman even became irritated about the Lord Chancellor's positing of *plicae* (folds) in matter as an alternative to the belief in the existence of the vacuum. In his *Novum Organum* (published in 1620 and studied by Beeckman in 1623), Bacon explains rarity and density by means of *plica materiae*, matter that can expand in space without producing a void and can contract without intersecting other parts of matter. Beeckman found this absurd.[66]

On the whole, Beeckman was confident, perhaps naively, in man's capacity to acquire knowledge of the natural world. He does not appear to endorse Calvin's theory regarding the tragic corruption of human understanding resulting from the Fall. In other ways, however, Beeckman's mechanical philosophy seems to have been influenced by his religious beliefs. He shared the Calvinist conception of the wide gulf separating God and nature, and the utter inability of matter to act on its own. In this context he often uses the phrase *unable not to*, in Latin *non potest non*. Double negation must come naturally to a believer in predestination, a doctrine that implies that humans are not in the position *not* to accept the grace of God.[67] Whereas the Remonstrants saw some room for the intervention of free will, the Counter-Remonstrants, including Beeckman, emphatically denied free will. Yet I do not want to overestimate the parallels between Beeckman's mechanical philosophy and his religious convictions. Half a century earlier, Perkins, in rejecting magical and devilish conceptions of nature, also emphasized that matter was totally dependent on God, yet Perkins was in no way a mechanical philosopher. There is no inevitable link between being a strict Calvinist and being a mechanical philosopher.

On the contrary, orthodox Calvinists were usually vehemently opposed to the new mechanical philosophy of nature. In the 1640s, Gisbertus Voetius— orthodox preacher, professor of theology, and rector of Utrecht University— bitterly fought the mechanical philosophy of René Descartes and his Dutch followers.[68] Acceptance of Descartes's absurd philosophy, which, according to Voetius, implied both atomism and Copernicanism, was a highway to atheism. The fact that Beeckman—born and raised in the Reformed Church and adhering to the same theological views as Voetius—was among the first in Holland to adopt the atomist philosophy that would inflame the Utrecht professor might seem incongruous. Beeckman, however, never had any trouble integrating his religious views with acceptance of atomism; his doubts about atomism were based on the internal logic of an atomistic account of motion, not on religious (or metaphysical) grounds. But Beeckman had developed his ideas in the years between 1607 and 1618, before the troubles in the Dutch Reformed Church and the young republic reached a climax. In 1618–19, Prince Maurice seized power, Oldenbarnevelt was tried and executed, and the Synod of Dordt convened to expel the Remonstrants and to redefine the Reformed creed. The orthodox leaders of the Reformed Church advanced Aristotelian philosophy beginning in 1619. Before this crisis, however, and certainly around 1610, even orthodox leaders had been much more open to new philosophical ideas. Whereas Beeckman saw no harm in combining his Counter-Remonstrant beliefs with atomism and mechanical philosophy, so the precocious medical student David Gorlaeus in his *Exercitationes philosophicae* (1612, published in 1620) tried to strengthen the case of the Remonstrants with an atomistic and vehemently anti-Aristotelian account of nature. Beeckman's formative years fell within this turbulent time, when all sorts of alliances were still possible; and the more restrictive atmosphere of the post-1618 years did not change his perspective.[69]

THE IMPACT OF RAMISM

Beeckman regularly observed that some of Gilbert's or Kepler's concepts were superfluous. Beeckman believed that a combination of mechanical and geometric properties of material objects eliminated the need for such concepts as *intelligentia, sympathia, sapientia in sole,* and *impetus.* For Beeckman, simplicity— understood as the use of as few explanatory concepts as possible—was an important criterion in assessing whether an explanation was acceptable. He formulated this concept at the beginning of the *Journal.* When he replaced

Scaliger's widely accepted explanation of the motion of the heavenly bodies through heavenly "intelligences" by his own explanation based on his new principle of inertia in 1612, he invoked a principle now usually known as Ockham's razor: "What can be done with a few means is said to have been done badly with many."[70]

Beeckman applied the same concept in his refutation of the idea of impetus. Presuming the existence of impetus does not really explain what happens when a stone continues to move after leaving the hand that has thrown it. Beeckman asked rhetorically, "Who can form an idea of what kind of thing that force is, how it keeps the stone moving, in what part of the stone it resides?"[71] But it is very easy to understand how nature works with the help of the concept of inertia: the principle that a given motion will continue forever unless it is impeded by an opposing force. As soon as one realizes that it is not movement as such, but only a change in motion that needs to be explained, the problem simply disappears. When confronted with a particular explanation for a natural phenomenon, Beeckman repeatedly used simplicity and intelligibility (taken visually) as criteria to assess whether the explanation was acceptable—or even necessary.

Simplicity and visually interpreted intelligibility in themselves do not constitute a natural philosophy comparable to Aristotelianism.[72] They represent a pattern of thinking unallied with any particular philosophy. In the fourteenth century, long before the dawn of the new science, the English nominalist philosopher William of Ockham (mentioned above) already employed simplicity as a methodological principle. In his *Summa of Logic* (*Summa totius logicae*, around 1324), Ockham wrote that it "is futile to do with more things that which can be done with fewer" (Frustra fit per plura quod potest fieri per pauciora).[73] Beeckman's formulation of his principle of simplicity ("Quod fieri potest per pauca, male dicitur fieri per plura") clearly echoes Ockham's maxim. The most direct source of this principle for Beeckman, however, is the sixteenth-century French philosopher and pedagogue Peter Ramus, or Pierre de la Ramée, whose philosophy is based on the ideals of simplicity and brevity. The mathematician Rudolph Snellius, Beeckman's teacher during his student days in Leiden, introduced him to Ramism. In later years, Beeckman also regularly refers to Ramus and his followers in his *Journal*. His library contained a number of Ramus's works. The ideas of this French philosopher and mathematician played a decisive part in the origin and development of Beeckman's mechanical philosophy of nature.

The Visual Thinking of Ramus

Peter (or Petrus) Ramus (1515–72) was the "enfant terrible" of the sixteenth-century French academic world.[74] He caused a great upheaval in conservative Paris in the 1540s by turning sharply against Aristotelian philosophy and presenting his own alternative to Aristotelian logic and dialectic. Despite the opposition of the Aristotelians, Ramus was able to attain a prominent position in the learned world. In 1545 he became rector of the Collège de Presles (part of the University of Paris) and was appointed regius professor in 1551 at what soon was to be the Collège de France. His life remained turbulent, however, particularly after his conversion to Calvinism around 1561. Beginning in 1568, he traveled extensively in Germany and Switzerland. Ramus, along with many other Huguenots, was killed in the St. Bartholomew's Day Massacre in August 1572.

Starting in the 1540s, Ramus wrote numerous textbooks for the trivium and the quadrivium. He was not a typical scholar, but first and foremost a teacher and educational reformer, a pedagogue whose views of science and philosophy were entirely determined by the way the subjects were taught. The classroom, a microcosm of external reality, constituted his world. Ramus's overarching pedagogical orientation had several important consequences. As a teacher, Ramus attached primary importance to the form in which knowledge was transmitted; he devoted much less attention to content. His textbooks were much more traditional with regard to content than the excitement they generated would lead one to suppose. Ramus regarded a clear presentation as essential and believed this meant that it was preferable to use the concrete to explain the abstract and to use practice to explain theory. Thus, a good mathematician not only reads Euclid carefully but also seeks advice from craftsmen to see how abstract mathematics works in practice. And a good logician not only teaches the principles of logic but also demonstrates how these principles are put to use in the arguments and reasoning of poets, philosophers, and orators.

These two examples are not arbitrarily chosen, because Ramus exercised great influence primarily in the areas of mathematics, logic, and dialectic (the latter two subjects were identical to him). His views on mathematics and the manner in which the subject should be taught are clearly expressed in his *Prooemium mathematicum* (*A Preface to Mathematics*) of 1567.[75] Ramus provided a survey of the history of mathematics, a defense of the value of the subject, and a general critique of the logical foundations of Euclid's *Elements*. Thus, he defended mathematics primarily by highlighting its practical significance. He carried his readers on an imaginary tour through the streets of Paris and showed them how mathematical knowledge was put to use everywhere by craftsmen, merchants, and state officials—mostly without their recognizing the mathematical nature of their work. Ramus liked to boast that there was no workshop in Paris that he had not investigated for practical applications of mathematical knowledge. Mathematics would have no reason to exist without such practical applications: the discipline was defined by its applications. Arithmetic was the skill of counting and figuring correctly, as in trade; geometry was the skill of measuring exactly, as in surveying. Even astronomy served only navigational needs; Ramus wanted to limit astronomy to creating tables for seafaring and carefully calculating in advance the positions of certain heavenly bodies. In keeping with this attitude, Ramus thought that mathematical propositions that had no apparent practical significance should be removed from textbooks.[76] He had no patience for speculative mathematics. According to Ramus, mathematicians who followed in the footsteps of Pythagoras and Plato, and assigned deeper symbolic or metaphysical meaning to mathematical figures, saw things in mathematics that were beyond the human capacity for observation. Indeed, Pythagoreans placed the object of mathematics outside observable reality ("in phantasticis et ab omnis sensu abstractis").[77]

Ramus did not advance the content of mathematics, but he believed that he could improve the order and presentation of existing mathematical knowledge. He wanted to eliminate superfluous proofs, senseless digressions, and unnatural ordering. The teacher's foremost task was the methodical ordering of the material.[78] Once that had been done well, Ramus argued, the propositions themselves would become much clearer, and they would produce their own proofs through the sequence in which they were presented.[79] In the correct sequence, propositions demonstrate their own validity. Thus, the truth of each proposition would immediately strike the observer when one followed the natural order. Ramus argued that this is possible because there is an agreement in principle between the order of things and the order of our understanding. Separate proofs are necessary only because our insight into the self-evident truth of propositions falls short; but a good, natural ordering makes such separate proofs superfluous. According to Ramus, the correct sequence proceeds from the general to the particular, and that which is derived should follow that from which it is derived. In geometry, the treatment of the curve thus should follow the treatment of the straight line, and in arithmetic, addition should precede subtraction.[80]

Beyond mathematics, Ramus influenced general ideas in the field of logic and dialectic. According to Ramus, traditional Aristotelian logic was too artificial and too complicated for pupils in the first years of their studies. For the thirteen- and fourteen-year-olds in his classroom, logic was only an introduction to philosophy and not an academic subject by itself. The students, therefore, profited from Ramus's quick and simple method of teaching the ground rules of thinking, which Aristotle did not provide.[81] Ramus thought he had created a better system.

Thus, Ramus essentially reduced all forms of reasoning to one fundamental form, scientific logic. Ramus's refusal to distinguish between logic as the teaching of reasoning in questions where certainty could be found and dialectic as the teaching of reasoning in matters in which only probable conclusions were possible was not revolutionary. This distinction between the domain of the certain and the domain of the merely probable had already become blurred in the later Middle Ages. But Ramus also wanted to eliminate the distinction between logic (or dialectic) and rhetoric, the art of persuasive reasoning. For Ramus, the difference between the rhetorical format of a speech and the logical arrangement of the arguments is not essential. He disputed the generally accepted idea that a speech delivered out loud and directed to a specific audience required a different structure than an essay written on paper and not directed toward any specific reader. If the natural disposition of the arguments serve as a guide—according to Ramus—then the conclusion would be so self-evident that it was no longer necessary to augment one's arguments with rhetorical tricks and devices.

Ramus maintained the traditional distinction in logic between *inventio*, the process of finding the arguments in support of a thesis, and *iudicium*, the process of arranging them in such an order that a conclusive and convincing demonstration would result. Ramus had little new to say about *inventio*. He adopted the system of *loci communes*, the enumeration of standard arguments, their storage in a convenient order and the techniques needed to retrieve these arguments from memory in its entirety. Ramus contributed new perspectives mostly to the *iudicium*, in which the speaker, after weighing the pros and cons, considered how the arguments raised could best be arranged in order. Earlier writers had taken into account the orator's personal insights or particular characteristics of the material under discussion, but Ramus separated the ordering of arguments from such "subjective" aspects of logical argument. He maintained that there was always a natural order, the most powerful and most convincing sequence, of arguments. It was not so much a question of weighing and interpreting arguments, as it was placing them in the correct, natural sequence. Instead of the interpretative *iudicium*, Ramus therefore preferred the more mechanical-sounding concepts of *disposition* and *collocatio*. The natural sequence was above all characterized by *claritas* and *distinctio*, lucidity and distinctiveness. Each part must be completely transparent to the mind and also clearly distinguished from related and nearby elements. Long before Descartes took clarity as a criterion of truth in his natural philosophy, the Ramists employed the notion that ideas had to be "clear and distinct" (clare et distincte) in their struggle against Aristotelian logic.[82]

Ramus's streamlining and simplification of traditional logic have often been misunderstood. Simplifying logic was not a matter of intellectual poverty but a program born of social ideals. By making logic more efficient and simpler, Ramus intended to make courses shorter and more effective. Ramus was the grandson of a struggling charcoal burner in Picardy and the son of a farmer who had died young. As a student, he worked as a domestic servant to other, wealthier students and had managed to scale the heights of the French academic system through hard work. "Labor omnia vincit" (Hard work conquers all), was Ramus's motto.[83] By reforming the university curriculum, making it more efficient, shorter, and more practical, he sought to reduce the costs of university education and thus open up higher learning to less privileged young men. This is why Ramus and his followers did away with the theoretical subtleties of the humanists, who seemed to be more concerned with explicating Aristotle's thoughts on logic than with preparing the students to use logic in further study or in practical life. As the Ramist Johannes Hachting, Franeker professor of logic, stated in 1627: "The logical precepts sufficient for a man active in public affairs ought in all respects to be brief, true, useful, and carefully arranged."[84] There is a definite shift in the Ramistic pedagogical program from theory to practice, from *verba* to *res* (from words to things), and here one can understand how Ramism prepared a new generation for the new philosophy that was gathering strength in the late sixteenth and early seventeenth centuries.[85]

Simplification often implies visualization. Indeed, Ramist logic was a visual kind of logic because of its emphasis on criteria that have a strong visual connotation (*dispositio, claritas, distinctio*). Ramus and his followers' use of graphic techniques in their textbooks confirms this point. In many ways, according to Ramus, *knowing* was equivalent to *seeing*. The Ramists preferred formal or quasi-geometric techniques, making use of outlines, tables, and above all diagrams in order to explain concepts and their correlations. Logical analysis of a concept or an entire academic discipline implied that the content would be incorporated into a detailed outline of dichotomies. Each concept was dissected and split into two symmetrical conceptual parts, each of which was separated in turn; the process continued until the fundamental parts of the original concept were discovered. One could then literally *see* the complete content of a concept, compressed onto one page of a book, in a tree of dichotomies. Many Ramists preferred to use this method to convey knowledge. The Swiss German Ramist Johann Thomas Freige, a contemporary of Ramus, published books consisting exclusively of such diagrams, without any linked text.[86] Freige's work embodies an extreme tendency of the "geometrization of logic," according to the Ramus scholar Walter Ong, a tendency that one also might describe as the "mechanization of thinking."[87]

Ramism in Leiden: Rudolph Snellius

For various reasons, scholars have long underestimated the importance of Ramism to the intellectual history of early modern Europe. First, the content of Ramus's logic is not very interesting, because it often simplifies Aristotle and his medieval commentators' logic. Second, many of Ramus's ideas can be traced to the work of his great predecessor, the Groningen humanist Rudolph Agricola, whose *De inventione dialectica* was written in 1479. Ramus's logic is a radical version of a more general humanist critique of Aristotelian logic. Finally, there are relatively few references to Ramus's thought in the philosophical literature between 1550 and 1650. This, however, is easy to explain. Ramus's ideas were particularly appealing to young students and their teachers. "Young people," one Ramist wrote to another in 1575, "read Ramus and the Ramists with the greatest eagerness in almost all the schools."[88] The simplified Ramist logic, claimed another, was "a godsend to young undergraduates, anxious to learn the art of disputation in the shortest possible time."[89] These adolescents later turned away from Ramism, although initially it had served them well. Yet this subsequent rejection of Ramism did not negate its influence. Ramus's logic influenced students in the most impressionable period of their intellectual development. "Once diffused in the classroom," the historian Owen Hannaway remarked, "Ramism seeped into the consciousness of Western Europe before it was too old to protest."[90]

Ramism achieved tremendous popularity between 1550 and 1650. During this period more than 800 editions of Ramus's and his most important collaborator Omer Talon's works were published, including more than 260 editions of

Ramus's *Dialectica* and more than 160 editions of Talon's *Rhetorica*.[91] Ramism spread throughout the Latin schools, the illustrious schools, and the smaller universities, especially in German-speaking areas, the Low Countries, and England (and via England, to Ireland and North America). Ramus's and his collaborators' social ideas appealed to many educators, as well as to the middle-class parents of students who flocked to the smaller institutions of higher learning springing up everywhere in northern and northwestern Europe in the late sixteenth and early seventeenth centuries. These students could not afford the erudite classical humanism of the Aristotelian professors in the more renowned universities, yet they still needed a modest degree of learning to apply for bureaucratic positions in the state and the church, which Ramism provided.[92]

Rudolph Snellius—mathematics professor at Leiden and Beeckman's teacher—was a very productive Ramist.[93] He was born in the small town of Oudewater between Rotterdam and Utrecht in 1546 and went to Germany at the age of fifteen to study Hebrew and mathematics at the University of Cologne. He also visited Jena, Wittenberg, and Heidelberg, but in 1565 he moved to Marburg, where he came under the spell of Ramism, probably during Ramus's visits to this Lutheran university in 1569 and 1570. Snellius stayed in Marburg for a number of years, studying at the university and teaching at the Paedagogium, a preparatory school for the university, led by the Ramist philosopher Lazarus Schoner. At the Paedagogium, Snellius's lessons covered "totam artium cyclopediam" (the complete cycle of the arts). One of his favorite pupils was the young Jacobus Arminius (1559–1609), the future leader of the Remonstrants in Holland and, like Snellius, a native from Oudewater.[94] Sometime after obtaining the degree of *magister artium*, Snellius went to Italy, where he visited Pisa and Rome and may have studied medicine. In 1575 he returned to Marburg to resume his teaching, but on hearing that the Spaniards had sacked his hometown of Oudewater in August 1575, he temporarily returned to the Low Countries to see what he could do for his family. After the fighting in the Low Countries had stopped, at least temporarily, in 1576, Snellius settled as a physician in Oudewater.

In the meantime the Dutch had created their own university in the city of Leiden, where they could train their own ministers, lawyers, and physicians. Leiden University officially opened on February 8, 1575, and one of the first students was Jacob Arminius, Snellius's former student and an ardent a supporter of Ramist philosophy, as his teacher was. He enrolled as a student of liberal arts in October 1576 and was soon—with a recommendation of Snellius—allowed to teach mathematics to his fellow students. Although Arminius eventually became a theologian and is best known for his disputes over predestination, he always had been very interested in mathematics. Petrus Bertius, who delivered Arminius's funeral oration in 1609, praised him for "his skill in Mathematics, and the other branches of philosophy in which his attainments were solid and profound."[95] Snellius himself also went to Leiden. He matriculated in October 1578 as a student in medicine, presumably not because he wanted to take additional courses, but because he wanted to profit from the privileges that were given to the students. He probably earned his living by housing boarding students and by private teaching. When Arminius left Leiden in 1581 to continue his study of theology in Geneva with Beza, Calvin's successor, Snellius took over his teaching obligations.[96] Snellius became "professor extraordinarius" of mathematics by 1581 and rose to the rank of full professor in 1601. He was rector of the university in 1607 and 1610. Snellius's son Willebrord succeeded him after his death in 1613.[97]

Snellius maintained his contacts with the learned world in Germany, where his books were still widely used in the 1590s.[98] Printers at Frankfurt and Herborn published Snellius's commentaries on the *Dialectica*, the *Arithmetica*, and the *Geometria* by Ramus and the *Rhetorica* by Talon. He also developed his own system under the title *Snellio-Ramaeum philosophiae syntagma* (Frankfurt, 1596). Snellius's short introduction to logic was reprinted six times, probably because it was used at the influential Calvinist college in Herborn. These books were published in Germany because opportunities to spread Ramus's ideas were much more limited in Leiden than in Germany. In Leiden, Snellius taught only mathematics; all the local philosophers were Aristotelians. Snellius did not actively oppose Aristotelian philosophy. He argued that he gladly would give up Ramus if someone could show him a better philosophy: the truth, nature itself—not Ramist ideas for their own sake—concerned him. Snellius reproached the Aristotelians for barely taking the trouble to read "The Philosopher's" texts and for failing to adequately distinguish Aristotle from his followers. Thus, Snellius was not a dyed-in-the-wool Ramist but rather a supporter of many-sided philosophical teaching, as he stated in his *Syntagma*: "No one can leave the university as a complete philosopher if he has not made a fundamental comparison and combination of the philosophies of Ramus and Aristotle."[99] In his books, Snellius refers to well-known Ramists such as Schoner and Freige and to semi-Ramists such as Philip Melanchthon, Rudolph Goclenius, and Cornelius Valerius. But Snellius's emphasis on the importance of practice as the test of true learning indicates that he really was a dedicated Ramist. "The origin and the beginning of philosophy lie in practical use [usus]; the purpose of philosophy is practical use; philosophy sets itself toward fruitful use for human existence."[100]

Two eminent humanists at Leiden—Justus Lipsius and Josephus Justus Scaliger—opposed Ramism. Prompted by Lipsius, in 1582 six students protested against the use of easy-to-digest compendia instead of the original texts of the ancient philosophers. The students argued that the university jeopardized its reputation if it allowed students to bypass the originals to study Cornelius Valerius's secondhand summaries. This incident clearly was directed against Snellius, who was teaching Valerius's books at that time. The Leiden senate agreed with the students and issued a decree that the professors should use Aristotle's texts exclusively when teaching philosophy.[101]

However, the humanists' success of was short-lived. In the 1590s the university boards had to issue new decrees to the same effect. For instance, Snellius's superiors instructed him to stop "reading Ramus" in 1591. They decreed that, henceforth, Snellius must limit himself to "Euclid or Aratus or Proclus or some other ancient author."[102] Yet the fact that the authorities had to repeat the decrees against the use of compendia and commentaries testifies to Ramist philosophy's continuing appeal to students and (at least some) professors, as well as the humanistic elite's determination to root out Ramism. Snellius remained popular among students: they carried him on their shoulders, ignoring the disdain of Snellius's more learned and conservative colleagues.[103] Snellius's influence on students extended beyond the mathematical realm: for example, Gerardus Vossius, later professor at Leiden and Amsterdam and a renowned philologist, had no aptitude for or interest in mathematics; yet, after he had returned to a moderate Aristotelianism in 1630, he recalled his enthusiasm for the dialectic of Ramus during his student years, between 1595 and 1598.[104] Vossius must have been referring to Snellius because the professor of philosophy at Leiden at that time taught only the *Organon* of Aristotle.[105]

Beeckman's Ramism

Isaac Beeckman also fell under the spell of Ramism in 1607 when he went to study in Leiden. He first learned some elementary mathematics from a distant relative, Jan van den Broecke, in Rotterdam. Beeckman probably was exposed to Ramism in Rotterdam, too, if Van den Broecke's 1609 textbook on arithmetic, geometry, and navigation can be viewed as a written record of his teaching. The book is organized, Van den Broecke repeatedly writes, "according to the doctrine of Ramus" (in Dutch, "na de leeringe van Ramus").[106] Van den

Broecke's book relies primarily on Ramus's work; he only briefly mentioned other mathematical practitioners, like Michael Coignet from Antwerp and Ludolph van Ceulen from Leiden.[107] After finishing his mathematical apprenticeship, Beeckman returned to Leiden, where he attended Rudolph Snellius's public lectures, including those on the *Optica* of Ramus.[108] Snellius also gave Beeckman individual study-help. He had approached Snellius to ask how he could best pursue the study of mathematics, despite having come to Leiden to study theology in 1607. As noted above, Snellius then gave Beeckman a list of books on various branches of mathematics.[109] Beeckman threw himself into his books with such enthusiasm that in the course of the winter he had to interrupt his studies temporarily because of exhaustion.

The reading list Snellius gave Beeckman had an unmistakable Ramist slant. Snellius divided mathematics into "pure" and "mixed" mathematics (*mathematica simplex* and *mathematica mista*):[110] Ramus's works were listed first in the pure mathematics category, followed by the usual works of Euclid, Hero, and Boethius. Under *mathematica mista*—that is mathematics applied to natural phenomena—Snellius included a mixture of important and less important names: Clavius, Commandinus, Copernicus, Giambattista della Porta, Euclid, Glareanus, Hermes Trismegistus, Hero, Oronce Finé, Pappus, Ptolemy, Regiomontanus, Stevin (for optics, not for mechanics), and Witelo.[111] Ramus's name appears only once, under *arithmetica mista*, but he is implicitly present elsewhere. Take, for example, optics, where Snellius recommended Euclid, Ptolemy, and Witelo. Snellius, though, did not refer Beeckman to Witelo's original work but to the 1572 edition published by Ramus's pupil Friedrich Reisner or Risner, the *Opticus Thesaurus* (which also included the work of Alhazen). Snellius used this book in his 1607 lectures (and undoubtedly Snellius had recommended Risner's book to Beeckman, because Beeckman referred to it at least once in the *Journal*).[112] With regard to Euclid's *Optics*, Snellius probably recommended the edition by Ramus's pupil Pena, published in 1557. In 1599 Snellius reprinted the propositions, though not the proofs and commentaries, of Pena's edition for use in his courses.[113]

After his studies in Leiden, Beeckman never renounced his sympathy for Ramism. Jeremias van Laren viewed Beeckman as a Ramist, judging by Beeckman's correspondence with Van Laren on logical and natural philosophical problems in 1613. Beeckman also referred to unnamed Ramist authors to strengthen his argument.[114] Furthermore, Beeckman's observations regarding the "unfruitfulness of philosophy" seem a reflection of his Ramist state of mind.[115] Likewise, Beeckman's 1618 dissertation exhibit characteristics of the

Ramist analysis of ideas: we are not sure about the theses themselves, because practically all of them are missing in the extant copy, but Beeckman definitely used the method of introducing dichotomies in analyzing the sciences in the oration he held on the occasion of the defense of his dissertation. In general, learning was divided into the categories of theology and philosophy; philosophy was further divided into mathematics and physics; physics in turn dealt with empty space and matter; and finally the analysis of matter entailed a division between form and motion (or rest, the absence of motion). It is a mechanical philosophy in a nutshell—a nutshell of Ramist making.[116]

Beeckman's role as a schoolteacher kept him from abandoning Ramism after graduating. He expressed Ramist ideas at the Latin schools in Utrecht, Rotterdam, and Dordrecht. Before Beeckman's appointment as vice-principal, Ramus's and Talon's books (the *Rhetorica* and *Dialectica*) were on the reading list in Utrecht, yet even Vergil's beautiful but useless *Aeneid* would have been replaced by Vergil's much more instructive *Georgics* had Beeckman been free to make such decisions.[117] And at Rotterdam, Beeckman expressed an interest in logical analysis and preferred dichotomies as well as methodologies that proceeded from the general to the particular. Moreover, he wanted his pupils in disputations to base their arguments on "the laws of the arts, which have been explained with so much dedication by the Ramists."[118] In the *Journal*, Beeckman adopted entire lines of argumentation from the *Logicae systema harmonicum* of the German encyclopedist and semi-Ramist Johann Heinrich Alsted. Beeckman also used the Ramist Freige's logical analyses of Cicero's orations.[119] In Dordrecht, Beeckman again used Alsted's work in teaching logic.[120] Beeckman did not like to teach abstract rules—such as his predecessor Gerard Borraeus had done. He had concluded—from reading Xenophon's *Kyroupaideia*—that the Greeks had not disputed *formaliter et logice* either but had followed their natural reason, just as in his own time the common folk sometimes debated more clearly than the pupils of the Latin school, despite their knowledge of logical laws and rules. Beeckman had great faith in the natural reason of common people and argued that sometimes the learned could be corrected by ordinary folk.[121] Thus, Beeckman was not in the least disturbed by the fact that his opinions, based as they were on close observation of the behavior of common people, sometimes deviated from those of Aristotle, as he wrote in a letter to Mersenne, dated October 1, 1629:

> I am not at all concerned about the truth of the axioms of the Aristotelians, since I know that the arts are taught by the people and approved by the people. The

people do not learn anything from the experts what the experts have not first observed in the people and, all being dispersed in the people, have collected badly. The people adjust their ears and voices to the musicians, but left to themselves, they retain what is good and convert the bad into the good, even though it unlearned it by frequent use.[122]

Ramus could not have expressed this better.

The 1637 auction catalog of Beeckman's books includes various titles by Ramus and commentaries on Ramus. Two copies of Ramus's *Grammatica latina*, two copies of the his *Grammatica graeca*, two copies of the *Commentaria Dounami in Dialecticam Rami*, and further Snellius's *Commentaria in Ethicam Rami*, Ramus's *Dialectica*, Johannes Piscator's commentary on the *Dialectica Rami*, the *Rhetorica* by Talon, Rudolph Agricola's *De inventione dialectica*, and both the *Logica* and the *Ethica* by Freige are listed in the "Miscellanei in Octavo."[123] Beeckman remained true to Ramism throughout his life, which was unusual for a scholar of his rank.

Ramism and Natural Philosophy

Ramism was much more than just a pedagogical philosophy for Beeckman. His mechanical philosophy is, to some extent, an application of Ramist principles. Walter Ong already suggested such a connection: "Ramist dialectic represented a drive toward thinking not only of the universe but of thought itself in terms of spatial models apprehended by sight. The world is thought of as an assemblage of the sort of things which vision apprehends—objects and surfaces."[124] Elsewhere he was even more explicit: "The whole process through Agricola and Ramus is thus of a piece with the tendencies discernible somewhat later, to deal with reality in a more visualist, observational, 'objective,' and, finally, mechanistic way."[125] However, Ong does not say how that link was established. But in his *Journal*, Beeckman shows how pedagogical reform could result in a reevaluation of natural philosophy.

First, Beeckman defends his way of proceeding with typically Ramist words and concepts at several points in the *Journal*, proving that he was thinking as a Ramist. For instance, Beeckman describes reasoning from principles to the ultimate components of the argument in 1620 as *compendiosius*, a strikingly Ramist expression (a compendium was something Aristotelians hated).[126] So even in his choice of his words, Beeckman is a transitional figure between the sixteenth-century schoolteacher and the seventeenth-century natural philosopher. Specific elements in Beeckman's natural philosophy also are directly related to his Ramist convictions. A prime example is his principle of inertia, which he almost certainly developed under the influence of Ramism. Ramist core intellectual values include clarity, simplicity, picturability, and the radical elimination of superfluous elements. Beeckman made calculated use of these values when he initially formulated and argued for his principle of inertia. At that time, Beeckman was corresponding with Jeremias van Laren, and cited Ockham's razor in defending his Ramist viewpoint on the relation between the logical concepts *individuum* and *species*.[127] As described above, Beeckman used the same principle to reject J. C. Scaliger's Aristotelian ideas that heavenly intelligences prompted the motion of heavenly bodies: "What can be done with a few means, is said to have been done badly with many."[128] According to this criterion, Beeckman's own concept of inertia, which required no separate moving force for the continuation of motion once it had been started, was preferable to that of Scaliger. From a later letter to Mersenne it can be inferred that he stated that he had conceived of his principle of inertia around 1609, when he was a full-time student in Leiden steeped in Ramist literature.[129] Thus, Ramism steered him away from Aristotelian mechanics and paved the way for the modern concept of inertia, so crucial for to the new science of the seventeenth century.

Beeckman's atomism or corpuscularism is also at least partly Ramist in origin. Ramus himself was neither an atomist nor a mechanical philosopher. Ramus, however, criticized aspects Aristotle's theory of matter. Perhaps Ramus's criticism helped open the way for a freer discussion of the natural philosophies of other classical authors, including the atomists. Yet Ramus's logic echoes that of the atomists, albeit in a metaphorical sense: he believed concepts could be dissected into their smallest constituent parts in order that they might subsequently—without the parts undergoing any change in meaning—be reassembled to form new concepts. At one point in the *Journal*, the connection between Ramus's logical atomism and Beeckman's physical atomism emerges: in a text by Galen about "atoms" and "forms," Beeckman noted a proposition in Ramus's logic, whereby a concept that applies to a single individual can also be a type concept. Beeckman proposes that as "individuum" is to "species," so "atom" is to "form," so that "form" consists of the total of the individual atoms, and their arrangement by rank or "configuratio."[130] This does not prove that Ramus's work was a direct source of physical atomism; classical authors as Hero, Lucretius, and Galen were more likely sources. Yet Beeckman's acceptance of Ramist logic increased his receptivity to such concepts. In any event, patterns of thought in Ramist logic harmonized well with theories of classical atomism.

The subject of Beeckman's inaugural address at Dordrecht in 1627 likewise links his natural philosophy and his Ramist background.[131] This oration, the only more or less systematic exposition of his general philosophy, is devoted to isoperimetric figures; a subject upon which Ramus wrote extensively in his *Geometria*.[132] Ramus wrote about isoperimetric figures in almost lyrical terms, finding them of particular significance for an understanding of practical life. In his *Scholae mathematicae*, Ramus expanded his hymn of praise to cover circular geometry in its entirety, citing Aristotle's authority: "Geometry provides the instrument of motion and rest. Without this instrument, nature can accomplish nothing, and the philosopher can understand nothing of the works of nature."[133] Beeckman warmly took these words to heart: without once mentioning Ramus's name, Beeckman's Dordrecht oration is a spirited defense of Ramus's opinions.[134]

Finally, Beeckman's preference for a visual representation of concepts over purely verbal and abstract principles also may be partially attributed to his Ramist education, because, for Ramists, knowing was basically seeing. I have written about this extensively in previous chapters, and so will not dwell on it here. Beeckman did not engage in the general practice of precise description that characterized much of Renaissance natural history. Although he studied medicine, he was not interested in anatomy; Vesalius is mentioned only once in his notebooks.[135] Yet Beeckman engaged in the "visual turn" of early modern science in a more general way, preferring explanatory mechanisms that can be visualized instead of purely verbal explanations. His Ramist background, no doubt, influenced to such preferences.

Ramism also hindered Beeckman's development: when Descartes introduced an infinitesimal approach to the falling-bodies problem in 1618, Beeckman had trouble understanding the concept. Beeckman's Ramist principles made it challenging for him to accept a concept—such as an infinitesimally small distance—that was an abstraction and had no counterpart in physical reality. He knew, in principle, that time and space were infinitesimally divisible, but the physical process of acceleration for him remained a discontinuous phenomenon: a falling object accelerates with little jerks, "met kleine hurtkens."[136] Descartes, on the other hand, had no such qualms and simply assumed that the acceleration of the object was a continuous process. He accepted Beeckman's physical representation only as an initial approximation, a useful tool in a purely mathematical argument. Beeckman, however, could not relinquish the direct link between mathematical concepts and their physical representation. Physical reality was always the norm for him; mathematics' role in the explanation of natural phenomena remained secondary. Ramism, therefore, may have kept Beeckman from exploring nascent mathematical concepts that might have deepened his understanding of motion.

CONCLUSION

Ramism, as I said earlier, is one of the most underappreciated movements in the history of Western philosophy. Key figures in the transformation of science and philosophy in the sixteenth and seventeenth centuries—even if they had been exposed to Ramist teaching—did not follow in Ramus's footsteps and as mature philosophers took their inspiration from other sources. Open adherence to Ramism was restricted to schoolteachers and pedagogues, mostly individuals of more limited learning and local renown. Beeckman's case, however, proves that Ramism could also influence the field of natural philosophy in a crucial way. As a schoolteacher and a craftsman, Beeckman used the intellectual tools of Ramism to translate the imperatives of the mechanical arts into the concepts of a new, mechanical philosophy. The Ramist emphasis on the visual representation of knowledge gave intellectual credibility to the craftsman's natural inclination to picture the world as an extension of the mechanical contraptions with which he worked. Religion's role is less clear. Although Beeckman's pronounced Calvinistic religious beliefs did not favor any particular natural philosophy, the Calvinist notion of the wide gap separating God and the world and the utter incapacity of matter to act on its own may have stimulated the translation of mechanical expertise into a mechanical philosophy of nature. But—in combination with the Ramist ideology of the schoolmaster and the mentality of the craftsman—even Beeckman's Calvinist beliefs provided a potent ingredient for a new view of nature and a new conception of what counted as an acceptable explanation of natural phenomena.

Alexandre Koyré once wrote that the transformation of natural philosophy in the seventeenth century was not simply the replacement of one set of theories by another. What happened was nothing less than a transformation of the fundamental categories of thought, "les cadres de l'intelligence elle-même."[137] We should assess Beeckman's contributions to the new science in this light. His mechanical philosophy revived ancient atomism, offered new explanations for musical harmonies, and introduced new concepts like inertia and the *homoge-*

nea physica, but at the same time this new philosophy fundamentally changed the criteria for evaluating explanations of natural phenomena. Many of Beeckman's ideas did not stand the test of time. Atomism foundered on its inherent contradictions, and the stress on pictorial representation proved untenable once infinitesimal mathematics and calculus transformed science. Yet the introduction of atomism and the use of picturability as a criterion for sound natural philosophy in the early stages of the scientific revolution were instrumental in preparing the ground for a science that was truly revolutionary.

CHAPTER SEVEN

Beeckman and the Scientific Revolution

Is the Ramist inspiration for Beeckman's mechanical philosophy important? Such an assessment ultimately depends on Beeckman's place in what historians still refer to as the scientific revolution of the seventeenth century. Was his an isolated case with minimal historical impact, or did his ideas play an important role in early modern natural philosophy? For a long time, historians minimized Beeckman's influence on the development of modern science. They regarded him as a minor philosopher without real influence, despite Beeckman's joint derivation of the law of falling bodies with Descartes. Beeckman did not do enough to promote his ideas, according to E. J. Dijksterhuis, who was otherwise sympathetic to the idea of publishing the *Journal* in its entirety. In his classic account of the rise of mechanism, *The Mechanization of the World Picture*, Dijksterhuis deplores Beeckman's lack of tenacity of purpose and powers of concentration required to systematize, finish, record, and publish his inquiries. As a result, he "did not advance science at all, or at least to a much smaller extent than [he] might have done."[1] Even in the case of Beeckman and Descartes's 1618 collaboration, Dijksterhuis claimed that Beeckman's ideas did not really form a link in the chain of development of seventeenth-century science. Beeckman's notebooks were valuable only "because they give the reader some notion of the scientific thought of a gifted man of the early seventeenth century."[2]

For Dijksterhuis, the history of science primarily was a chain of theories, of which the science of mechanics was the most important. In this history of mechanics, Beeckman's influence was limited by his failure to publish his derivation of the law of falling bodies. Yet this assessment does not do justice to

Beeckman's contribution to the development of natural philosophy. Our conception of science has broadened considerably over the past half century. Mechanics no longer takes center stage and "science" in general is now viewed as a set of practices that are only partially articulated, and not simply as a set of theories. This broader concept of early modern science has consequences for a definitive assessment of Beeckman's legacy.

First of all, one should deemphasize, as Beeckman did, his work on the law of falling bodies. He focused on many other disciplines, including music, chemistry, optics, and meteorology. Second, disseminating his views on these topics took different forms. Dijksterhuis presumes that knowledge is transmitted only through books and underestimates the role of oral conversation and written correspondence. Some scholars had difficulty acquiring books, and at times the scholarly community took years to integrate new ideas they encountered in books. Exchanging letters was often a faster and more direct way of informing others of one's work than publishing a book. One should therefore not underestimate the extent to which Beeckman influenced the course of seventeenth-century natural philosophy through his correspondence and personal contacts with Descartes, Gassendi, and Mersenne—in fact, writing a letter to Mersenne was the early modern equivalent of publishing an article in a scientific journal. Finally, analyzing Beeckman's *Journal* also directly and indirectly demonstrates the importance of artisanal knowledge in the rise of modern science. Beeckman repeatedly used his workshop experience to make sense of natural phenomena in the *Journal*, while Ramism—rooted in the world of practical knowledge—gave him confidence that this was a sensible thing to do. These ideas coalesce in the notion of picturability introduced above to capture the essence of Beeckman's approach. This notion places Beeckman at the heart of the scientific revolution, while establishing how Beeckman's work fits in the general culture of the Dutch Republic in the early seventeenth century.

BEECKMAN, GASSENDI, AND MERSENNE

Other scholars have challenged Dijksterhuis's assessment that Beeckman had limited influence on his contemporaries. In his Mersenne biography, Robert Lenoble wrote, "It seems that the first initiator of a great number of the ideas of the seventeenth century was the modest and taciturn Beeckman."[3] Many well-known discoveries of seventeenth-century science were, indeed, first expressed in Beeckman's notebooks: for example, Beeckman noted the correlation between the height of a column of water and the flow velocity of water pouring through an opening at the bottom of the column thirty years before Torricelli published on the subject.[4] Of course, Beeckman did not publish his discovery, and so Torricelli had to "discover" it on his own.

Beeckman, though, was not a recluse: he shared his speculations with others and was not as reticent as Lenoble supposed. Beeckman spoke about many of his ideas to all who expressed genuine interest in his activities. Most of those visitors subsequently made little use of what they learned from Beeckman. For example, Colvius—the Walloon preacher in Dordrecht with whom Beeckman made occasional meteorological and astronomical observations—was probably well informed about Beeckman's natural philosophy. Colvius knew many Dutch scholars and might have shared Beeckman's theories with them, but he did not. This explains why Constantijn Huygens never mentioned Beeckman's name, although Huygens was very curious about developments in natural philosophy in the Dutch Republic. On the other hand, Pierre Gassendi, who visited Beeckman in 1629, spoke with others about what he learned, and he definitely used Beeckman's ideas in developing his own system of thought.

When Gassendi traveled through the Dutch Republic, he met twice with Beeckman, who provided his French guest with an overview of his mechanical philosophy. In addition, Beeckman gave Gassendi a copy of his dissertation (or at least the revolutionary corollaries) as a gift. In a letter to Peiresc, Gassendi calls Beeckman "the best philosopher I have met so far."[5] Nevertheless, Gassendi did not obtain a specific mechanistic solution for a specific natural philosophical problem from his conversations with Beeckman. The two of them discussed concrete subjects and exchanged ideas in their infrequent correspondence, such as on the parhelia observed at Rome in 1629. Their conversations about natural philosophy in general had a greater impact, because after meeting Beeckman—a devout Protestant and a convinced atomist—Gassendi believed that atomism could be reconciled with Christian faith. In the preceding years, Beeckman had carefully mediated a position halfway between the ancient atomists, who were naturalists and believed in spontaneous creation through a *fortuitus concursus*, and biblical creationists, who believed that God had personally created each and every living being. Through his notion of the "primordia"—the building blocks of nature created by God from which only specific things could be constructed—Beeckman in a sense had Christianized ancient atomism. He had placed a Christian foundation under Lucretius's naturalistic explanation of the generation of natural things. This position appealed to Gassendi, and he adopted it as a starting point for his own ideas about the generation of living beings.[6] His meeting with Beeckman encouraged him to pursue his plan for developing a Christianized version of ancient atomism as an acceptable and attractive alternative to Aristotelian natural philosophy.[7]

Of course, there were considerable differences between the atomism of Beeckman and that of Gassendi. For example, Gassendi ascribed to his atoms a *vis motrix*, an internal capacity for motion that is strongly reminiscent of the active principles and occult qualities in Renaissance philosophies, or the old scholastic concept of impetus. Beeckman, a strict mechanist, would not have accepted such a concept. On the other hand, Gassendi's principle of inertia, published in 1642, echoes Beeckman's. Although Gassendi may have heard others discuss this idea, Beeckman definitely discussed his principle of inertia with Gassendi in 1629. So, although Gassendi never read Beeckman's *Journal*, Beeckman clearly influenced his work.[8]

Beeckman's influence was even more substantial on those who actually read his *Journal*. The Dutchman rarely gave others access to the *Journal*. Indeed, in December 1626 he expressly disapproved of the publication of his notes. The Dutch Republic was going through difficult times, having just lost the city of Breda to the Spanish and still mourning the death of Prince Maurice in 1625. Beeckman was afraid of being criticized for possible errors he had made and also fearful that others might use his ideas for malign purposes, including passing them on to the Spaniards. Therefore, he resolved to write notes in Dutch only, because the country's enemies could not benefit from them (in fact, Beeckman continued to write in Latin as well as Dutch). By showing his notes to a few trusted friends, however, Beeckman hoped to ensure that any useful ideas in his notebooks might be circulated among the trusted. Three would suffice.[9] And, indeed, only Descartes, Mersenne, and his favorite pupil, Martinus Hortensius, were allowed to consult Beeckman's notebooks.[10]

Of these three men, Beeckman's *Journal* had the least impact on Hortensius. Hortensius was a very ambitious young man, most interested in astronomical questions and related topics such as navigation and optics. Beeckman had been Hortensius's tutor and had certainly stimulated the young man's interest in astronomy. He also had convinced him that the Copernican system was correct. Then, in 1628, Beeckman introduced Hortensius to Philip van Lansbergen. Lansbergen had been a good friend of Beeckman's father and had helped Beeckman when he studied medicine in Middelburg. In the intervening years, Lansbergen had become a prolific writer on astronomical subjects and had acquired a reputation as a staunch defender of the Copernican system.[11] Thus, prompted by Beeckman's recommendation, Lansbergen took Hortensius under his wing and taught him the finer points of astronomy. Hortensius, in turn, helped

Lansbergen carry out some research and translate his book on the Copernican system. When Hortensius became professor of mathematics in 1634 at the newly founded Athenaeum illustre in Amsterdam and began teaching optics and astronomy, he was the first professor in Holland to openly support the Copernican system.[12] It was only then that Beeckman showed Hortensius his notebooks.

Mersenne made more use of his access to the *Journal*, although he read it (most likely only some of the Latin notes) in 1630 primarily to test Descartes's assertions about the true nature of his relationship with Beeckman. Beeckman probably allowed Mersenne to see his notes when the French friar came to visit him and informed him of the misinformation Descartes had spread about him. It is quite possible that Mersenne saw only the passages related to that topic. Afterward, however, Mersenne corresponded with Beeckman on a regular basis. They discussed problems in musical theory in great detail. Mersenne integrated some of Beeckman's ideas into his books on musical theory, where he cites Beeckman's work. For example, in his *Harmonicorum libri* of 1636, Mersenne uses Beeckman's explanation of the phenomenon of a single string producing different tones, inserting a quote from a 1633 letter from Beeckman. Mersenne also reported the opinion of the "acutissimus philosophus I. Beeckmannus," this time quoting a 1631 letter in his explanation of the "phenomenon" of the *media quies* (the point of rest between two movements).[13] While Mersenne ignored many of Beeckman's ideas—for instance, his corpuscular theory of matter—Mersenne, through his published works, disseminated some of Beeckman's concepts to the wider community of mathematicians and natural philosophers in the early seventeenth century.

BEECKMAN AND DESCARTES

Beeckman's relationship with Descartes had an even greater impact. The standard view has long been that Beeckman's influence on Descartes was real but minimal. Descartes omitted mention of Beeckman in the *Discourse on the Method* (1637), Descartes's account of his intellectual development. In that little book, Descartes portrayed the genesis of his entire philosophy as arising from the troubling visions that he had had in Germany in 1619 and 1620—a year *after* his meeting with Beeckman. Later Cartesian scholars acknowledged that Descartes had been extremely complimentary to Beeckman in his letters during the early months of 1619, but they denied that Beeckman had influenced the content of Descartes's philosophy. "The encounter of Descartes and Beeckman

in 1618 has an extreme importance in the intellectual history the father of Cartesianism," the French philosopher Étienne Souriau acknowledged in 1938.
"But it is a simple occasional cause, and leaves no trace on Cartesian philosophy."[14] Beeckman simply had spurred Descartes to his philosophical mission,
without having influenced in any way the philosophy that he subsequently
developed.

However, after the discovery of the *Journal* in 1905, Cornelis de Waard's publication of several articles on its contents, and the 1939 publication of the *Journal*'s first volume, scholars could no longer deny that Beeckman indeed had
influenced Descartes in significant ways. So Alexandre Koyré in 1939 concluded:
"In effect, one now realizes that Beeckman fully deserves the title of *vir ingeniosissimus* with which Descartes honored him; and, what is more, he henceforth
appears as a link of primary importance in the history of the evolution of scientific thought. Finally, his influence on Descartes seems to have been much more
profound than has been supposed up to now."[15] Koyré contends that his analysis of the cooperation between Beeckman and Descartes in 1618, particularly on
the motion of falling bodies, provides convincing proof of this statement. Yet,
although Koyré considered Beeckman's influence to be substantial, he also
claims that Descartes misunderstood Beeckman, implying that, after a couple
of years, Descartes had to reinvent what Beeckman had taught him before—
thereby again diminishing Beeckman's influence.

Descartes's presentation of his intellectual development in his *Discourse* is
the primary reason scholars largely underestimated Beeckman's contributions
to Descartes's thought. John Schuster already chronicled the misleading nature
of Descartes's account. Descartes's metaphysics and epistemology are *not* the
logical consequence of his visions of 1619 and 1620; and the roots of his mechanistic natural philosophy can be traced back to his earlier meeting with Beeckman.
In 1618 Descartes was mainly interested in natural philosophy, and he began to
address the metaphysical and methodological foundations of his mechanical philosophy only in the 1630s. He believed he had to change course because there
were some problems with the elaboration of his natural philosophy. Indeed,
Beeckman had nothing to offer Descartes with respect to metaphysics, but
Beeckman initially surely inspired Descartes to build his natural philosophy.
Descartes's study with Beeckman, so Schuster contends, "set the tone of his
career as a philosopher of nature. In effect, Descartes served a second natural
philosophical apprenticeship with Beeckman, an apprenticeship which forti-
fied him with a new vision of the aim and content of natural philosophy, and
which displaced the Scholastic vision purveyed at La Flèche."[16] Beeckman

provided Descartes with a primitive model of the physical-mathematical phi-losophy and a number of problems to which the new method could be applied, such as the motion of falling bodies, the resonance of vibrating strings, and the hydrostatic paradox. These small research projects served as preliminary exercises for Descartes's work in optics, which led him in the 1620s to the inde-pendent discovery of the law of refraction, a discovery that he viewed as a mag-nificent demonstration of the fruitfulness of his natural philosophy. According to Schuster, Descartes's physical-mathematical philosophy at that point basi-cally amounted to the rewriting of propositions from mechanics in terms of the corpuscular matter theory that he had learned from Beeckman. Inspired by Beeckman, Descartes constructed "micromechanical" or corpuscular models to explain the observable phenomena of mechanics.[17]

This assessment, although basically correct, is too simplistic. It does not ac-count for the sophistication of Beeckman's philosophy, much less that of Des-cartes. Schuster, indeed, acknowledges that there were differences between Beeckman and Descartes. Beeckman's mechanics was very much in the tradition of the *Quaestiones mechanicae*, whereas Descartes's was more mathematical, Archimedean.[18] For example, a remarkable formulation of force as "the incli-nation to motion" already appears in Descartes's "Hydrostatic manuscript" of 1618, while it is absent in Beeckman's notebooks. Also, Descartes's later search for metaphysical and theological guarantees for his mechanistic principles, an important element of his natural philosophical program, is absent from Beeck-man's natural philosophy.

Yet one should not put too much stress on the importance of micromechani-cal models in Beeckman's mature philosophy of nature. Beeckman's 1627 inau-gural oration at Dordrecht presents his physical-mathematical philosophy without referring to these micromechanical models; instead, he outlines the properties of isoperimetric figures, by discussing phenomena that take place on the macromechanical level. For Beeckman, physical-mathematical reason-ing in general was the explanation of natural phenomena using a combination of physical and mathematical (i.e., geometric) properties of material bodies. In many cases, the corpuscular structure of matter provided the physical ele-ment in that reasoning, but not necessarily. At other times, the principle of in-ertia sufficed. Nor does the corpuscular philosophy play any direct role in Beeckman's proof of the inverse proportionality of the length of a string and the pitch of the tone produced (although corpuscular mechanisms do play an important role in other aspects of his musical theory). Tiny particles occupy a prominent place in many of his explanations, but not always; sometimes Beeckman does not need them to explain natural phenomena. Therefore, corpuscular notions are not essential to a physico-mathematical proof as such. The physical-mathematical philosophy was a general method of explanation and logically independent of the specific theory of matter that was embraced. Physico-mathematics does not presuppose a corpuscular theory of matter and, therefore, is not identical with constructing micromechanical models. Beeckman's contribution to Descartes's thought thus went deeper than the utility of devising such models.[19]

Arthur recently elaborated on Beeckman's influence on Descartes. He claims that in 1618 Descartes had taken a mature "ontology of motion" from Beeckman. Unlike Koyré, Arthur contends that "there was no misunderstanding between them about the physics of fall in 1618, and that they conceived motion in fundamentally the same way: in terms of God's conserving a body's motion, together with a certain force of motion proportional to the speed and magnitude of the body, and shorn of any connotations of internal cause."[20] Basically, Beeckman's *Journal* notes lay out this ontology of motion *before his meeting with Descartes*; there is no evidence that Descartes arrived at these conclusions on his own. Also, Descartes did not deny that these ideas were correct in 1630; he merely claimed that Beeckman was not their author. For this claim, Descartes adduced two reasons. First, he simply claimed that he initially had come up with these ideas himself. For example, when Descartes discusses Beeckman's theory of motion with Mersenne in 1630, he said: "He supposes, as I do [supponit, ut ego], that what has once begun to move continues to move of its own accord," an implicit denial of priority that must rank, according to Arthur, "as one of the most disingenuous ever made in the history of physics."[21] Descartes's second argument was that the true value of ideas depends on the system in which they are put to use, and because his own philosophy is much more systematic than the fragmented *Journal*, Beeckman should stop claiming these ideas as his own. Now the first argument is simply a lie, and the second not much more than a clever sophism, and so the conclusion must be that much of what is normally attributed to Descartes's genius should instead be attributed to Beeckman. Referring to the nine principles of motion on which Beeckman and Descartes agreed in 1618, Arthur reaches the same conclusion: "To the extent that Descartes learnt the nine principles of motion from Beeckman, acknowledged that he did, and then subsequently and very publicly denied he had learned anything from Beeckman, used his influence to suppress knowledge of his debt to his former mentor, and took sole credit for their shared views, one can fairly say that a significant part of what subsequently became known as Cartesian natural philosophy was plagiarized from Beeckman."[22]

Of course, not everything in Descartes's mature natural philosophy or even his theory of motion is derived from Beeckman. The *Journal* does not contain the idea that inertia applies only to linear motion, nor does it mention the notion of an instantaneous tendency to motion. Furthermore, the context in which Descartes worked out his ideas in the 1620s was quite different from that in which Beeckman lived and worked.[23] Even in the early 1630s, when both Descartes and Beeckman lived in the same country, the public Descartes had in mind differed completely from the kind of people Beeckman was addressing. Descartes knew very well that in order to get his philosophy accepted at the schools in France, he had to provide it with a sound metaphysical foundation. In the Dutch Republic, however, natural philosophers avoided such questions to keep them from getting embroiled in theological disputes, like the one that had raged through the young republic in the 1610s. Consequently, Dutch thinkers often remained stuck in an eclectic, patchwork philosophy because they understood the benefits of avoiding "metaphysical foundations" or theological consequences of their ideas. The fragmented *Journal* is therefore highly representative of Dutch intellectual culture in the seventeenth century, while Descartes still had an eye on the intellectual culture prevalent in his native France.

Nevertheless, Descartes's debt to Beeckman is so obvious, that Descartes's furious reaction, upon learning what Beeckman had written to Mersenne in 1629, appears more understandable. The vehemence with which Descartes in 1630 tried to distance himself from Beeckman simply indicates that his natural philosophical work in progress still resembled that of the rector of Dordrecht much more than the French philosopher was ever prepared to admit. When Descartes met Beeckman again in 1628 and 1629, he soon learned that Beeckman had further developed his ideas along the lines Descartes himself had been working along since 1618, and that Beeckman was considering publishing some of his ideas.[24] Beeckman showed Descartes the *Journal*, now bound in calf and with new marginal notes to the entries to facilitate easier reference. In some of his recent notes, Beeckman also alluded to the possibility of publishing his ideas on the corpuscular mechanisms responsible for the behavior of heavenly objects to demonstrate that his explanations were much better than Kepler's. So Descartes encountered Beeckman at a time when the latter was seriously contemplating publishing the same ideas that Descartes was treating in the manuscript that later became *Le monde*.

All this more than likely unnerved Descartes.[25] He understood Beeckman's limitations well enough to realize that the Dutchman would not succeed in writing a truly systematic treatment of natural philosophy, including its foundations. Even if Beeckman were to publish first, a more dispassionate philosopher would have concluded that most people would see the real differences between the two enterprises on the publication of Descartes's more systematic treatment. Thus, Descartes would still receive the honor he believed he deserved. But Descartes could not abide the thought that anyone else could explain natural phenomena in a way that echoed his own. Descartes, therefore, succeeded in discouraging Beeckman from carrying out his plan to publish his ideas, just like he launched, at about the same time, an offensive to discredit his former collaborator Ferrier.[26] Beeckman cancelled his plans to publish his ideas and devoted himself to other interests, such as grinding lenses. And when Descartes's star began to rise in the 1630s and 1640s, no one suggested that the great French philosopher owed a considerable debt to the reserved rector of Dordrecht.[27]

But Descartes's dependence on Beeckman did not go entirely unnoticed in the seventeenth century. When Descartes's *Principia philosophiae* and Beeckman's *Centuria* were both published in 1644, the Nijmegen preacher and antiquarian John Smith noted the remarkable resemblance between the two books. On August 22, 1644, Smith wrote to Constantijn Huygens: "Recently in Zutphen thanks to your kindness I have read Descartes's remarks on the magnet, but later I read the *Centuria meditationum mathematico-physicarum* of the Dordrecht rector Isaac Beeckman (written in 1628, but only published this year), in which he shows under number 36, 77, 81, and 83 that Descartes was not the first to think of those [magnetic] corpuscula."[28] Huygens did not respond to this information, and it has taken historians until late into the twentieth century to conclude that Descartes's corpuscular natural philosophy (fig. 8), even taking into account all the differences, is indebted in many respects to Beeckman's ideas.

PICTURABILITY AND THE RISE OF A
MECHANICAL PHILOSOPHY

Not only has Beeckman's influence been far greater than Descartes scholars have acknowledged, but his special way of treating natural philosophical problems was also much more representative of a general trend than has been recognized until now. Historians of science working in the Koyré-Dijksterhuis

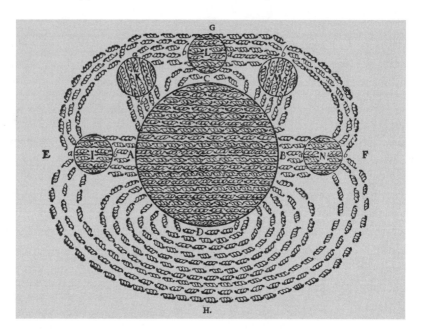

Figure 8. Descartes's explanation of magnetism (from Descartes's *Principia philosophiae*, 1644)

tradition, with its heavy stress on conceptual analysis and its glorification of the mathematization of science in the seventeenth century, did not appreciate Beeckman's special style. It is easy to understand why: Beeckman's lack of conceptual rigor, his tendency to favor the visual over the abstract, and his fondness for devising corpuscular mechanisms to explain phenomena did not align well with these historians' intellectual standards. In *The Mechanization of the World Picture*, for example, Dijksterhuis constructed the concept of mechanization so that it was a priori impossible for someone like Beeckman to play an important role in the story of the scientific revolution.

In the epilogue to his magnum opus, Dijksterhuis distinguishes four meanings of the words "mechanization" or "mechanistic thought."[29] The most common meaning rests on etymological considerations: the understanding of the world as a machine. Many scholars in the seventeenth century regularly made that comparison, saying that the world was like a mechanical clock. Nonetheless, according to Dijksterhuis, this comparison was peripheral to mechanistic thinking in science, as expressed in the work of Galileo, Huygens, and Newton. The idea of a Creator acting with a specific purpose, putting the machine together, is altogether lacking in their work. Popular accounts of the new science perhaps invoked the comparison with the clock, but, as Dijksterhuis said, "science as such never paid the slightest heed either to a supra-mundane Maker of the Universe or to an extramundane object which He might desire to pursue by His creation."[30] Mathematicians and natural philosophers did not consider the purpose for which the machine was assembled, which is essential to the comparison of the world to a machine. They were concerned how the machine worked, how the gearwheels were powered, and how the cogs meshed together, but the machine's ultimate purpose remained beyond the pale. Yet anyone who analyzes a machine without considering the purpose for which it was created is not really treating it as a machine but only as an unspecified mechanical system, which according to Dijksterhuis is something completely different.

Dijksterhuis also discarded two other possible interpretations of mechanization. The first involves defining mechanization in reference to the lifelessness of both nature and the machine. Like a machine, an external force must activate nature; it cannot prompt its own actions. But this places one in a difficult position: even Aristotelianism, using such a parameter, could be viewed as mechanistic.[31] But one runs into other difficulties if one adopts the other alternative and attempts to define mechanization by stressing the visual element in mechanical explanations. Looking to machines and mechanisms built by skilled craftsmen, one can see how they work; and this also applies to mechanisms that explain natural phenomena. These mechanisms became the paradigm of intelligibility in natural philosophy *because* it was easy to picture them. Dijksterhuis acknowledges that in the history of natural philosophy, certainly in England, natural philosophers often proposed mechanisms to represent the physical reality behind the phenomena. However, they did not limit themselves to existing mechanisms and mechanical tools such as pulleys, gear wheels, and levers, but also invented mechanisms that no craftsman had ever constructed. Dijksterhuis referred to such "mechanisms" as the collision of atoms and the vortex motion, particularly dear to the hearts of Cartesians. "However, when people also began to apply the term 'mechanistic' to the mental pictures operating with such motions and to speak of a mechanical model when a natural phenomenon was explained with their aid, the connection with the basic concept of instrument was considerably lessened."[32]

An added disadvantage of this interpretation is that by focusing on the possibility of mentally picturing natural phenomena, this term *mechanical* comes to participate in the vagueness and the subjectiveness inherent in the notion of picturability. According to Dijksterhuis, the concept of picturability was simply too vague to be of any use. Besides, what was considered to be picturable changed over time. This is true to a certain degree. Pictures can represent reality in different ways. Mechanical philosophers implicitly understood pictures always to be *mimetic* representations of reality, but pictures could be used in other ways, too. In the early seventeenth century, Robert Fludd believed that his *emblematic* or *symbolic* pictures offered an ocular demonstration of the close relations of microcosm and macrocosm; and Kepler's and Mersenne's denial that such pictures had any demonstrative power really offended Fludd.[33] In the late seventeenth century, Huygens and Leibniz rejected Newton's idea of universal gravitation as being nonpicturable and therefore not admissible to mechanical philosophy; yet half a century later, natural philosophers could easily imagine how two bodies acted upon each other at a distance.[34] Universal gravitation had become as real to them as atoms and subtle matter had been to Beeckman and Descartes.

Dijksterhuis reasoned that the idea of picturability, given its pliability, could not possibly function as the defining characteristic of mechanical philosophy. However, the reliance of mechanical philosophers on the science of mechanics as the basis of natural philosophy remained constant over time. Mechanics, that is the mathematical analysis of motion, was the core of Newton's and his opponents' philosophy of nature. Consequently, the true meaning of mechanization is simply the application of the science of mechanics to the description and explanation of natural phenomena. Because mechanics was primarily a *mathematical* theory of motion, mechanization also became identified with the mathematization of science over time. When Albert Einstein cut the last tie to picturability by giving up the idea of ether in the early twentieth century, the identification of mechanization and mathematization was complete. Thus, Dijksterhuis concluded his book with the classic statement that the mechanization of the world picture during the transition from ancient to classical science "meant the introduction of a description of nature with the aid of the mathematical concepts of classical mechanics; it marks the beginning of the mathematization of science, which continues at an ever-increasing pace in the twentieth century."[35]

If the history of science were nothing more than a rational reconstruction of the development of scientific thought, there would be no reason to argue with Dijksterhuis. Newton's concept of force acting at a distance—in a way that no craftsman could replicate—was the crowning achievement of mechanistic natural philosophy in the seventeenth century and laid the foundations for further developments in the science of mechanics. Indeed, Newtonian science appears to make it impossible to seek the essence of mechanization in anything other than the increasing mathematization of the natural sciences. Yet such reasoning contains the defects of unmitigated finalism, pretending that the outcome of a development retroactively sums up the entire course of development and explains it. But often history is not so linear; and the outcome of a development sometimes may be the unintended consequence of initiatives that were undertaken for entirely different reasons. In any case, one should not presuppose the end result if one wants to understand how people in the early seventeenth century thought about the process; or if one wants to arrive at an explanation for the rise of mechanistic natural philosophy, which began long before Newton. If one wants to know how the mechanization of the worldview actually developed, then studying how a man like Beeckman understood fundamental principles of the natural sciences becomes a window into the living history of science.

By downplaying the role of visual imagery and pictorial representation in the scientific revolution, Dijksterhuis deviated from the general trend among his contemporaries; other historians of science in the 1940s and 1950s, such as Herbert Butterfield and Koyré, viewed the scientific naturalism of the fifteenth and sixteenth centuries as a significant factor that contributed to the revitalization of science in the seventeenth century.[36] These historians studied the work of Leonardo da Vinci, Vesalius, Albrecht Dürer, and other artists with "scientific" inclinations, while others analyzed the work of a mathematician and astronomer like Galileo, who definitely had a feeling for art. The next generation of historians of science tended to overlook this interaction between art and science; in the 1970s and 1980s, the philosophy of science and the social study of science were much more prominent.[37] The situation has shifted again, however, and today's historians of science explicitly incorporate the visual aspects of science into their work. But today the scope of the discussion is much broader than it was half a century ago. Today's historians analyze the role of the visual in early modern science as a more general phenomenon; they study the inherently pictorial nature of much of the new science of the seventeenth century as well as the specific pictures and illustrations used by people in that era.[38]

Here, Beeckman's work reemerges. Pictures as such do not seem to play a very important role in his work, yet his philosophy is highly visual, in the sense that Beeckman can understand something only insofar as he can visualize it. Beeckman could get rid of the verbalism inherent in Aristotelian philosophy of nature only by making mental pictures. This is what I mean by the notion of picturability. In Dijksterhuis's view, picturability was a misleading sideshow at best and not at the core of mechanization; but it definitively was a cornerstone of Beeckman's physical-mathematical philosophy. Descartes's mechanical philosophy also is loaded with visual imagery. As Christophe Lüthy has shown recently, Descartes's natural philosophy is the first natural philosophy that is illustrated with figures, diagrams, and scenes from everyday life.[39] Whereas Renaissance textbooks of natural philosophy lack pictures, Descartes realized that most people would regard the picture as an integral, constructive, and even the most easily remembered part of the argument. The image of the magnet with the surrounding stream of magnetic particles is probably the best-known illustration of the essence of Descartes's mechanical philosophy (fig. 8).

Of course, the increasing use of illustrations by philosophers accompanied advances in printing technology in the sixteenth and seventeenth centuries. Yet the use of such illustrations also reflects the changing nature of natural philosophy. Descartes's natural philosophy was more visual, certainly in comparison to the abstract, logical theories that dominate the Aristotelian philosophy of nature.

Picturability, or *Anschaulichkeit*, is the common element in Beeckman's and Descartes's natural philosophy. Such illustrations were essential to their mechanical philosophies. However, the picturability resided not in the drawings and the engravings themselves but in the concepts and criteria of intelligibility these philosophers adopted. This is also true for other mechanical philosophers of this generation, including Gassendi, Walter Charleton, Kenelm Digby, and Hobbes, who believed that they could replace the Aristotelian or scholastic philosophy of nature with clearer, more comprehensible systems. This intelligibility resulted from the fact that they "made use of sensible principles such as matter, Locall Motion, Magnitude, figure, situation &c. so that when hee speakes, it is possible to understand what he means," as the mechanical philosopher William Petty remarked in 1647.[40] More precisely, when a mechanical philosopher explains natural phenomena in terms of the arrangement in space, shape, and motion of individual particles of matter, one *sees* what he means. The mechanical philosophers were confident that seeing had a persuasive power that could not be matched by anything else.

For this reason, mechanical philosophers in general had high expectations of what they could achieve using the new optical instruments, such as the telescope and especially the microscope. For us, such speculations about the micromechanisms that explained natural phenomena are simply speculations; but for many early seventeenth-century mechanical philosophers, the possibility of actually seeing these micromechanisms seemed within the realm of possibility. Beeckman was very interested in both the telescope and the microscope (see the copy of Drebbel's letter with the drawing of a microscope) and devoted hours of hard work to learn the craft of lens grinding in the 1630s: apparently, he strongly believed in the potential of the new optical instrument to prove the mechanical philosophy right.[41]

There is also a remarkable parallel between Beeckman's stress on picturability of natural philosophy and the unprecedented realism of Dutch art in the seventeenth century. Svetlana Alpers has argued that a descriptive impulse characterized Dutch art in this period.[42] Dutch paintings in general do not tell a story, she claims, as Italian paintings often do, but they describe a scene, be it a landscape, a bowl of fruit, or a young woman reading a letter. She contends that a Dutch painting is not "read," like an Italian painting, but "seen." Dutch paintings address the eye and not the mind. In their depiction of everyday scenes, Dutch paintings are often dazzling in their display of craftsmanship and radiate a pleasure in carefully depicting silk dresses, decaying fruit, or light shining through the clouds. This descriptive impulse predates the Reformation, and Calvinism—with its stress on the Word—did not mitigate it. According to Alpers, the descriptive impulse of Dutch painters runs parallel to the empirical and experimental tendencies in the world of Dutch science and learning in the seventeenth century. She points out that there are indeed several interesting connections between the history of art and science in this period. Constantijn Huygens, the secretary to several stadholders and a man of fine taste, was an art lover *and* also someone with a keen interest in Baconian philosophy and Drebbel's inventions, particularly his microscope. Furthermore, Alpers notes that several Dutch painters had a keen interest in optics, specifically in the theory of perspective and the use of the camera obscura and the magnifying glass.

Alpers also cites Beeckman to support her theory—because he was an early reader of Bacon and because of his keen sense of observation. The attentive eye with which Beeckman records scenes in his diary from his personal and intellectual life evidently captivated Alpers. After relating some of Beeckman's observations—lumps of sugar sucking up the liquid around them, the movements of the clouds, the disposition of panes in the window just over his seat in the church, the flicker of a candle, the fact that books are normally higher than they are wide, and even the dissection of the body of his brother Jacob after he had died from consumption—Alpers concludes:

> In recording his observations, Beeckman compiles the bits and pieces of a program for Dutch art. Dutch artists shared a passion for visual attentiveness with

this kind of experimenter and savant. Indeed, their works attend to many of the same things that he does: Ruisdael's billowing clouds lifted high over the land; the swaying of the candle and the curled pages of piled books captured for our eyes in Leiden still-life paintings; the cadavers depicted by portraitists, to whom death presents itself in the form of a surgeon's lesson in anatomical observation.[43]

Of course, this does not mean that Beeckman deep down in his heart was a painter like Jacob van Ruisdael of Gerrit Dou or Rembrandt. Beeckman never showed any interest in the fine arts and regularly complained of poor eyesight. It means that the natural philosopher and the painter were both driven by this common "passion for visual attentiveness" that according to Alpers was so peculiar to Dutch culture in the seventeenth century.

Art historians have severely criticized Alpers's thesis for its one-sidedness and its disregard of other traditions and influences.[44] Alpers intended her book to serve as a critique of the slightly older tradition of seeing hidden meanings in apparently purely descriptive paintings, but there is no denying that many paintings are loaded with moral messages and references to emblems and common sayings—Alpers herself draws our attention to some painters' predilection for proverbs and commonplaces. Utterly realistic depictions of landscapes or tiny insects can have religious or moral meanings. Furthermore, many Dutch painters—not only those who actually had been in Rome—adhered to the Italian style because there was a market for such work. Although the market for simple landscapes and genre pieces was larger, historical topics and mythological themes were also popular in Holland. Finally, the so-called realistic pictures—from the interior of a church to a wide-open landscape—often were in fact sophisticated constructions of scenes that would never have been possible in reality. What is so striking about many Dutch paintings is not the realism of the picture but rather, as Mariët Westermann has pointed out, the "reality effect," the conscious effort of the painter to create a lifelike representation, to imitate nature as best as he could even though the composition as such was unrealistic. Verisimilitude, the conscious effort to create a lifelike impression, seems to be more characteristic of Dutch painting in the seventeenth century than realism.[45]

Painters who lived near Beeckman aspired to create this reality effect in their pictures as well. Ambrosius Bosschaert (1573–1621) was a Middelburg painter well known for his flower pieces. Beeckman may have been acquainted with the man and his work. Like Beeckman's father, Bosschaert was a Flemish immigrant and became a well-respected craftsman in his new place of residence. At least six times he served as dean of the guild of St. Luke, the guild of the painters and the art dealers. One can easily imagine Beeckman and Bosschaert meeting at the home of one of the Middelburg regents, where Beeckman was installing a fountain in the garden filled with tulips and other flowers, which the regent then commissioned Bosschaert to paint. The realism in Bosschaert's paintings is striking. Each flower is meticulously depicted and insects and frogs are added to enliven the whole; there are even a few dewdrops that suggest that the flowers have just been picked from the garden. But Bosschaert's flower pieces do not represent an actual bouquet of flowers, because it contains blossoming flowers from different seasons. Moreover, the vase is put in a stone arch window that was unusual in Dutch architecture and the background is a distinctly non-Dutch, mountainous landscape. The realism of the painting is just an illusion. Instead of viewing this picture as a lifelike depiction of an actual bouquet, one should value it as a collection of rare flowers and other natural objects (shells, insects, amphibia). A flower piece brings together rare objects and sets them in a relationship like objects in a cabinet of curiosities. And yet, despite its artificiality, the reality effect is overwhelming.[46]

So, after all, there is a kernel of truth in Alpers's thesis about the peculiarly realistic impulse of Dutch art. For although Dutch painters used multiple strategies to suggest reality more than to depict it as they found it, and although they used these strategies in genre pictures and still lifes as well as in pictures that have manifest or hidden moral or religious meanings, Westermann can still conclude that "Dutch artists seem to have been more concerned with verisimilitude than their colleagues in almost any other Western culture."[47]

Anyhow, the apparent realism of Dutch paintings should inspire historians of science to take a closer look at the connections between art and science in the Netherlands in the seventeenth century, and especially at Beeckman's *Journal*. The meticulous representation of material things in Dutch still lifes developed in a society that was obsessed with things.[48] It was an early capitalist society in which things came to be valued primarily as commodities. Of course, this was not a purely Dutch phenomenon. The new money economy engulfed most of Europe; Italy, southern Germany, and Flanders had experienced the transition earlier. But no country embodied the new economy as fully as the Dutch Republic. Whereas most other parts of Europe were still dominated by an agrarian economy, in which political power and social rank were determined by landed property, in Holland industry and commerce—that is, the production of and the trade in material goods of whatever kind—became much more important, although wealthy merchants in time began to imitate the aristocratic lifestyle and acquired manorial rights. The new mentality led to a predilection for paintings with detailed and realistic representations of material things, whether drinking bowls on the table, carpets on the floor, a map on the wall, or a lady's fine dress. The new obsession with things also drove the remarkable trade in shells, tulip bulbs, and various exotic objects from the Indies. The tulip craze of the early seventeenth century, ending in the crash of 1637, when prices of bulbs fell dramatically and the owners of these bulbs lost lots of money in just a few months, is a famous, albeit rather extreme example of this obsession with collecting wondrous things.[49]

Those who could not afford such expensive items, or wanted to preserve the sight of the exquisite flowers, could commission a painter to make a flower portrait or could buy pictures representing flowers, shells, and other collectables. For example, the aforementioned Ambrosius Bosschaert began by painting individual "portraits" of exotic plants, probably commissioned by wealthy plant growers in Middelburg. Growing exotic plants had become fashionable in upper- and middle-class circles in Middelburg around 1610, and several botanists and plant collectors, some of whom traded information with the Leiden botanist Carolus Clusius, lived in Middelburg.[50] Caspar Parduyn—whose father owned a large garden in which he cultivated rare plants acquired from Clusius, some of them immortalized in Bosschaert's paintings—was one of Isaac Beeckman's closest friends in Middelburg.[51] Realism was essential in these pictures, because buyers were not interested in flowers in general, but in specific flowers, such as the checkered lily (*Frittilaria meleagris* L.), the orange flower-scented flag (*Iris plicata*), and the *Narcissus tortuosis*. The obsession with material things is certainly not the only explanation for the remarkable realism of seventeenth-century Dutch painting, but it adds an interesting perspective to the phenomenon.

Beeckman did not take part in the obsession with material things that was so important in seventeenth-century Dutch society. There are no references in the *Journal* to shells or other exotic objects, to flowers in his garden, or to paintings on the walls of his house. He did not participate in the tulip craze. He explicitly stated that he was rather careless about his clothes. Actually, he appears to have been totally unrepresentative of the material culture of the Dutch Republic. And as far as his philosophy of nature is concerned, the things that most interested him, the atoms that constitute matter, were and remained invisible. Finally his ideas about free-falling bodies, musical harmonies, and colliding atoms required an element of mathematical reasoning that completely falls outside the scope of the theories of Alpers and others. In Beeckman's natural philosophy, picturability does not imply producing a realistic picture of the components of nature. Beeckman even declined to picture the shapes of the four kinds of atoms he distinguished.

Yet, on a deeper level, Beeckman's pursuits mesh well with the picture of a middle-class society obsessed with material things and characterized by a strongly visual attitude toward nature and the world. His philosophy was in essence the philosophy of someone who was accustomed to manipulating nature rather than contemplating it; someone who knew that the bodily encounter with matter produced knowledge as much as studying ideas in books. Beeckman's philosophy of nature is indeed primarily a philosophy of *things*, not of verbal or mathematical *relations*. And in his own way, he exhibited the descriptive impulse that art historians find so characteristic of seventeenth-century Dutch paintings. Beeckman objected to the concept of impetus because he could not mentally *picture* what kind of object this impetus was and where it resided in the moving object. However, the impulse to picture the world in his philosophy goes deeper than the merely descriptive. For Beeckman, it was enough to see how everything worked with his mind's eye, so he hardly ever felt the need to actually picture the mechanisms of nature by making precise drawings of them. The purpose of the rather crude drawings in the *Journal* is therefore only to *assist* the mind in getting a clear mental picture of the phenomenon at issue. What mattered to Beeckman was that his theory was capable of being portrayed graphically as opposed to actually making the drawing. Picturability served as a criterion for demarcating reasonable explanations from fuzzy speculations; it was not the goal of natural philosophy to picture nature.

Picturability in natural philosophy of course had its problems. In time, it occurred to Beeckman that postulating invisible mechanisms analogous to the mechanisms that we see in the world around us actually meant shifting rather than solving the problem. Postulating that objects that rebound after collision consist of particles that have the same qualities as the larger object, and thus have the same problems, does not really explain the original phenomenon. Yet this fact did not induce Beeckman to give up picturability as a criterion, for to him it was a matter of principle. Picturability was not some vague notion, applied without awareness of its inherent limitations, or something that historians of science can identify retrospectively as a theme running through Beeckman's thought. Picturability was the essence of natural philosophical thinking for Beeckman. And the problems inherent in a strict adherence to picturability are certainly no reason to deny this generation of mechanical philosophers their place in history.

Beeckman was the oldest of his generation of mechanical philosophers. Whereas Descartes and Gassendi could build on Beeckman's theories; and Hobbes and Charleton could build on Descartes's and Gassendi's, Beeckman had to devise his own mechanical philosophy. As shown in the previous chapters, he devised his notion of picturability with the help of the Ramist philosophy to which he was exposed at least from the days he was studying with Rudolph Snellius at Leiden. This preference for pictorial strategies and representations fitted well with his artisanal background and his Reformed faith. Given Ramus's appreciation for artisanal knowledge and his conversion to Calvinism, both elements had been "baked into" Ramist philosophy. Yet it is highly unlikely that without his immersion in Ramist thought Beeckman would have given the pictorial such a key position in his natural philosophy. Despite his craft background, Beeckman needed a consciously formulated philosophy like Ramism to enable him to fundamentally break with Aristotelianism and construct a mechanical philosophy of nature. Before the idea could arise that it might be useful to apply mechanics and mathematics to the study of natural philosophical problems, he had to develop a conception of reality that made the world suitable to that application. On that basis, one can argue that a more fundamental kind of mathematization of nature preceded the application of mathematical concepts, that the essence of mathematization was not the formulation of mathematical correlations between specific phenomena, but the logically much more fundamental process of clearing the ground and making the world fit for the application of mathematics. In Beeckman's case, it was Ramism that performed this duty. Even if one views this history of the mechanical philosophy in the light of the final results achieved by Isaac Newton, the emphasis on picturability was a vital part of the process. Indeed, picturability was a necessary intermediate phase, an indispensable link in the chain that reached its culmination in Newton. The emphasis on pictorial representation in the early phase of the scientific revolution was a necessary step on the road to the fully fledged mathematization of science in the second phase, culminating in Newton's *Mathematical Principles of Natural Philosophy* (1687).

By the time Newton published his *Principia*, Beeckman had been dead for fifty years. Beeckman would have been surprised or even shocked to see how mechanical philosophy had been transformed during those fifty years. The mathematics employed would have been beyond his comprehension and the notion of gravity would have appeared to him—as it did to Christiaan Huygens and others—as surrendering to just the kind of philosophy he had done his best to defeat. Still, Beeckman also would have seen what fine results could arise from this new kind of mechanical philosophy, and he would have noticed that throughout Europe there was still this common idea that, in one way or another, the world had to be explained in mechanical terms. How different had it been in the early decades of the seventeenth century. When Beeckman set out to explain the world according to principles that a common-sense craftsman and well-educated mathematician like himself could fully understand, there was no common ground for all those who were dissatisfied with the old-fashioned Aristotelian or scholastic notions. There was an enormous diversity of alternative positions, and only gradually out of this chaotic situation did a new and fruitful tradition emerge through the works of Kepler, Galileo, and Descartes. They defined what problems were interesting, what methods were adequate, and what kind of solutions would count as satisfactory. Newton's accomplishments, for all their genius, had been built on the work of these predecessors. Newton saw further—as he said—because he was standing on the shoulders of giants. Beeckman, however, lived at the dawn of this new science; there was no tradition for him to follow. To use a metaphor more fitting for a Dutchman: he had to sail alone in uncharted waters.

Notes

INTRODUCTION

1. Historians of science nowadays are reluctant, and for good reason, to use the term *scientific revolution*. The singular concept of *science* is an anachronism, and a process that took almost two centuries to reach its climax (in the work of Isaac Newton) hardly qualifies as a revolution. There was no revolution *in* science; at most the science that emerged was the result *of* a revolution. It is therefore for purely pragmatic reasons that historians continue to use the term *scientific revolution* as shorthand for all the astonishing things that happened in the investigation of nature in the sixteenth and seventeenth centuries. An excellent survey of the different interpretations of this episode in Western history is provided by H. Floris Cohen, *The Scientific Revolution: A Historiographical Inquiry* (Chicago: University of Chicago Press, 1994). For some recent surveys of the period itself, see the bibliographical essay at the end of this volume.

2. Alexandre Koyré, *Etudes galiléennes* (Paris: Herman, 1939) (an English translation by John Mepham was published as *Galileo Studies* in 1978). Herbert Butterfield, *The Origins of Modern Science: 1300–1800* (London: Bell and Hyman, 1949), and E. J. Dijksterhuis, *The Mechanization of the World Picture*, trans. C. Dikshoorn (Oxford: Oxford University Press, 1961), belong to the same tradition.

3. Paolo Rossi, *Philosophy, Technology, and the Arts in the Early Modern Era*, trans. Salvator Attanasio (New York: Harper & Row, 1970 [1962]). See also Edgar Zilsel, *The Social Origins of Modern Science* (Dordrecht: Kluwer Academic Publishers, 2000); the original articles were published in the 1940s.

4. Rossi, *Philosophy, Technology, and the Arts*, xi.

5. A. Rupert Hall, "The Scholar and the Craftsman in the Scientific Revolution," in M. Clagett, ed., *Critical Problems in the History of Science* (Madison: University of Wisconsin Press, 1959), 3–23, quotations on p. 16.

6. Pamela H. Smith, *The Body of the Artisan: Art and Experience in the Scientific Revolution* (Chicago: University of Chicago Press, 2004). Elsewhere Smith nicely summarizes the characteristics of artisanal knowledge: "It was disciplined by years of practice, was precise, cumulative, experimental, investigative, demonstrated (visually and practically), collaborative and an example of distributed cognition. It was largely acquired through observation and imitation, rather than through texts, because written descriptions leave out too much information and would not develop flexibility in responding to always-changing environments. It responded to particular and fluid situations. It relied upon external demonstration within a public setting and was dependent on a community of practitioners. Finally, it resided in

and was 'proven' by objects." Pamela H. Smith, "In a Sixteenth-Century Goldsmith's Workshop," in Lissa Roberts, Simon Schaffer, and Peter Dear, eds., *The Mindful Hand: Inquiry and Invention from the Late Renaissance to Early Industrialization* (Amsterdam: Edita, 2007), 33–57, quotation on p. 44. Harold J. Cook, *Matters of Exchange: Commerce, Medicine, and Science in the Dutch Golden Age* (New Haven: Yale University Press, 2007), provides another example of the expanding reach of the social interpretation of early modern science. Cook claims that "the intellectual activities we call science emerged from the ways of knowing most highly valued by the merchant-rulers of urban Europe" (p. 40). The book focuses on the interactions between such merchants and mainly physicians and practitioners of natural history, in both the Dutch Republic and it its overseas territories.

7. The definition of a *liminal figure* is taken from Smith, *The Body of the Artisan*, 26. These liminal figures may not have been central to the traditional story of the scientific revolution, Smith argues, but they are "central to understanding the complexity of the past, and they assist us in transcending received narratives about the development of modernity (most especially the positivist history of the rise of modern science" (p. 19).

8. The distinction between the scholar and the craftsman, taken for granted by a previous generation of historians of science such as A. Rupert Hall, is now considered to be problematic. In Roberts, Schaffer, and Dear, *The Mindful Hand*, this distinction, as well as other ingrained dichotomies like theory-practice, head-hand, and science-technology are fundamentally criticized. The editors claim that instead of anachronistically speaking about science and technology, one should refer to "natural inquiry" and "invention." They view what is usually called the history of science and technology as the history of "a single, hybrid affair in which the work of the head and of the hand formed a complex whole" (pp. ix–x). They also argue that the historical recovery of a mixed regime of the mindful hand is all the more difficult to recover as long as historians effortlessly accept the old distinctions (p. xiii). One should indeed be aware that some people at the time attempted to turn these socially determined and normative distinctions into self-evident facts of life, which they are not. However, many people in the early modern period sincerely believed that these distinctions between manual labor and scientific knowledge were real, and they acted upon this belief, thereby *making* them real. Rejecting these distinctions out of hand may therefore be as anachronistic as using them.

9. Actually, the Middelburg bookseller Jutting bought the manuscripts during a public auction and then passed them on to the Provincial Library.

10. *Journal tenu par Isaac Beeckman de 1604 à 1637, publié avec une introduction et des notes par C. de Waard*, 4 vols. (The Hague: Martinus Nijhoff, 1939–53). One cannot have but the greatest admiration for the meticulous care with which De Waard edited the notebooks, but one has to be aware that—as he stated in his introductions—he left out some passages (mainly meteorological tables) and rearranged some of the notes (among others the notes on lens grinding and two nosographies). Therefore, the oration Beeckman delivered on becoming rector of the Latin school at Dordrecht appears in the fourth volume and not in the third volume, which contains notes from that period. De Waard also felt free to interpolate the texts where he thought this would be helpful. It has become customary to refer to Beeckman's notebooks as his *Journal*, although Beeckman did not use this term; it was purely De Waard's idea to call them (in French) a "journal." Beeckman started his notebooks as a collection of *Loci communes* and also used the title *Meditata mea*. The fourth volume of De Waard's edition differs from the preceding volumes in the sense that it is mainly documen-

tary. It contains, aside from a few parts taken from Beeckman's notebooks, mostly archival documents relating to his activities over the years. After a fire destroyed most of the archives in Middelburg, during the German attack on the city on May 17, 1940, these printed sources are practically all we have left to reconstruct Beeckman's life (*Journal*, 2:457). The original manuscript was damaged too (mainly water damage) but has been restored. It is now stored in the Zeeuwse Bibliotheek in Middelburg (ms. no. 6471).

11. Cornelis de Waard, "Sur les règles du choc du corps," in *Correspondance du P. Marin Mersenne religieux minime*, 17 vols. (Paris: Beauchesne, 1932–88), 2:632–44; Koyré, *Études galiléennes*; R. Hooykaas, "Science and Religion in the Seventeenth Century: Isaac Beeckman, 1588–1637," *Free University Quarterly* 1 (1951): 169–83; H. Floris Cohen, *Quantifying Music: The Science of Music at the First Stage of the Scientific Revolution, 1580–1650* (Dordrecht: Reidel, 1984); M. J. van Lieburg, "Isaac Beeckman (1588–1637) and His Diary-Notes on William Harvey's Theory on Blood Circulation (1633–1634)," *Janus* 69 (1982): 161–84; H. H. Kubbinga, *L'histoire du concept de "molecule,"* 3 vols. (Paris: Springer, 2002), translated as *The Molecularization of the World-Picture, or The Rise of the Universum Arausiacum* (Groningen: University of Groningen Press, 2009).

CHAPTER 1: THE MAKING OF A NATURAL PHILOSOPHER, 1588–1619

1. There is no recent history of the city of Middelburg. Still relevant is W. S. Unger, *Geschiedenis van Middelburg in omtrek* (Middelburg: Altorffer, 1954). See also P. J. Meertens, *Letterkundig leven in Zeeland in de zestiende en de eerste helft der zeventiende eeuw*, Verhandelingen der Nederlandse Akademie van Wetenschappen, Afdeeling Letterkunde, Nieuwe Reeks, Deel 48, no. 1 (Amsterdam: Noord-Hollandsche Uitgevers Maatschappij, 1943), 8–13, 18–19. For a more recent evaluation of the scholarly life in Middelburg around 1600, see Klaas van Berkel, "The City of Middelburg, Cradle of the Telescope," in Albert van Helden et al., eds., *The Origins of the Telescope* (Amsterdam: KNAW Press, 2010), 45–72. On the Dutch Revolt, see Jonathan I. Israel, *The Dutch Republic: Its Rise, Greatness, and Fall, 1477–1806* (Oxford: Clarendon Press, 1995), and other literature mentioned in the bibliographical essay at the end of this volume.

2. On the number of immigrants entering Middelburg, see N. W. Posthumus, *De geschiedenis van de Leidsche lakenindustrie*, 3 vols. (The Hague: Martinus Nijhoff, 1908–39), 2:69–75, 890–92. For a trustworthy estimate of the population of Middelburg, see Peter Priester, *Geschiedenis van de Zeeuwse landbouw circa 1600–1910*, A.A.G. Bijdragen, 37 (Wageningen: Afdeling Agrarische Geschiedenis, 1998), 52–58, 481. In the older literature, figures are highly exaggerated. On the importance of the immigration from the South, see J. G. C. A. Briels, *De Zuidnederlandse immigratie 1572–1630* (Haarlem: Fibula-Van Dishoeck, 1978). Oscar Gelderblom, *Zuid-Nederlandse kooplieden en de opkomst van de Amsterdamse stapelmarkt (1578–1630)* (Hilversum: Verloren, 2000), offers a more recent reevaluation of the impact of the southern immigration on the Dutch Republic.

3. Decline set in shortly after 1600, when the influx of immigrants from the south dropped dramatically and commercial activities moved to northern cities like Amsterdam and Rotterdam. Middelburg around 1590 collected as much custom duties as Amsterdam, but by 1620 Amsterdam collected three times as much as Middelburg. See Priester, *Geschiedenis van de Zeeuwse landbouw*, 72; Victor Enthoven, *Zeeland en de opkomst van de Republiek: Handel en strijd in de Scheldedelta ca. 1550–1621* (Leiden: Luctor et Victor, 1996).

4. Biographical information, unless otherwise noted, is taken from Cornelis de Waard, "Vie de l'auteur," in *Journal*, 1:i–xxiv. De Waard also wrote the entry on Beeckman in *Nieuw*

Nederlandsch Biografisch Woordenboek, vol. 8, cols. 84–88, as well as those on Abraham Beeckman Sr., Abraham Beeckman Jr., and Jacob Beeckman (ibid., cols. 84, 88–89).

5. De Waard did not do any archival research in Turnhout. This gap was filled by E. van Autenboer, "Emigranten uit Turnhout: De familie Beeckman(s)," *Vlaamse Stam: Tijdschrift voor Familiegeschiedenis* 4 (1968): 144–50. One of Beeckman's genealogical notes in his *Journal* states that during his exile in London, his grandfather Hendrick Beeckman refused to meet an old friend from Italy, Chiapin Vitellus, because Vitellus had become known for mocking religion. Vitellus, so Hendrick Beeckman reportedly said, "might poke fun at me that I put myself in this trouble for the sake of religion" (Vitellio soude met my gecken dat ic om de religie my in dit ongemack gesteken hebbe) (Genealogical note in Beeckman's notebooks, f. 98r; *Journal*, 1:236, note). Elsewhere, Beeckman explicitly mentions that his grandfather had fled "because of his religion" (om de religie) (Genealogical note in Beeckman's notebooks, f. 49v). This is essential because many immigrants left the southern Netherlands for economic reasons. On the Dutch colony in London, see Johannes Lindeboom, *Austin Friars: Geschiedenis van de Nederlandse Hervormde Gemeente te Londen, 1550–1950* (The Hague: Martinus Nijhoff, 1950); Heinz Schilling, *Niederländische Exulanten im 16. Jahrhundert: Ihre Stellung im Sozialgefüge und im religiösen Leben deutscher und englischer Städte* (Gütersloh: Mohn, 1972), 45–51.

6. There is a contradiction between what De Waard said on the place where Beeckman lived after his marriage in his "Vie de l'auteur," *Journal*, 1:iii, and his note to *Journal*, 4:2–3. The latter is more likely correct. During the research for volume 4 of the *Journal*, De Waard reconsidered statements he had made in the first volume.

7. Beeckman wrote in 1618: "The first three sons were born at the Livestock Market, diagonally opposite to the house on the corner of the 's Gravenstraat, in a small little house. The others in the Hoogstraat diagonally opposite to the City Shed in The Two Roosters" (De 3 eerste sonen sijn geboren op de beestemarckt, noes over het hoeckhuys van de s'grave-strate in een klijn huysken. De reste in de hooghstrate noes over de stadtsschuere in de twee hanen) (Genealogical note in Beeckman's notebooks, f. 49r). Both houses are still standing; the latter is now called "The Two Fighting Cocks" (De Twee Kemphanen).

8. *Journal*, 4:9.

9. The candles made by the elder Beeckman seem to have been of excellent quality. The minister and astronomer Philip van Lansbergen, who lived in Goes (twenty kilometers away), regularly sent a box with the skipper of a trading barge to Middelburg, to be filled by Beeckman with his best candles, as if they were—to make a modern comparison—fine cigars. Lansbergen to Beeckman, November 16, 1604, Zeeuwse Bibliotheek (Provincial Library of Zeeland), ms. 5078 (cf. ms. 5079). (See note 14, below.)

10. *Journal*, 2:179.

11. Pieter de La Rue, *Geletterd Zeeland* (Middelburg: Callenfels, 1742), 8.

12. *Journal*, 1:iv.

13. On Biesius, see J. G. C. A. Briels, "Zuidnederlandse onderwijskrachten in Noordnederland 1570–1630. Een bijdrage tot de kennis van het schoolwezen in de Republiek," *Archief voor de Geschiedenis van de Katholieke Kerk in Nederland* 14 (1972): 89–169, 277–98; ibid., 15 (1973): 103–49, 263–97. On the Arnemuiden Latin school, see Hendrik Wilhelm Fortgens, "De Latijnsche scholen te Veere en Arnemuiden," *Archief Zeeuwsch Genootschap 1944–1945*, 51–73. On the Latin school system in general, see Fortgens, *Schola latina: Uit het verleden van ons voorbereidend hoger onderwijs* (Zwolle: Tjeenk Willink, 1958).

14. Official documents concerning the conflict between Beeckman and the Middelburg ministers have been collected by De Waard in the *Journal*, 4:8–16. He also mentions two collections of letters by and to Beeckman in the Municipal Archives of Vlissingen (1:v, n. 3; 4:8, n. 9, 16, n. 6), without using them or publishing abstracts in the fourth volume of the *Journal*. They are now kept in the manuscript collection of the Zeeuwse Bibliotheek, ms. 5065–67, 5069–91. Concerning the problem of whether children of Roman Catholic parents could be baptized in the Reformed Church, see A. Th. van Deursen, *Bavianen en Slijkgeuzen: Kerk en kerkvolk ten tijde van Maurits en Oldenbarnevelt* (Assen: Van Gorcum, 1974), 135–39; Hendrik Jan Olthuis, *De doopspraktijk der Gereformeerde Kerken in Nederland (1568–1816)* (Utrecht: Ruys, 1908), 78–96.

15. In a letter dated May 17, 1600 (Zeeuwse Bibliotheek, ms. 5070), Lansbergen mentions preceding letters of March 30, April 6, and April 14, but these letters have not been preserved. On Lansbergen (1561–1632), who was a minister in Goes since 1589, see *Nieuw Nederlandsch Biografisch Woordenboek*, vol. 2, cols. 775–82. Lansbergen, who originated from Ghent, had lived in London from 1566 to 1579 and presumably befriended Abraham Beeckman in that period. A few years later, the preacher became well known as an astronomer and defender of the Copernican system. See Rienk Vermij, *The Calvinist Copernicans: The Reception of the New Astronomy in the Dutch Republic, 1575–1750* (Amsterdam: Edita, 2002), 73–97.

16. See, for instance, the undated letter of the minister Herman Faukelius to Hans de Swaef, a supporter of Beeckman, Zeeuwse Bibliotheek, ms. 5072.

17. Genealogical note in Beeckman's notebooks, f. 49r.

18. De Waard gathered this from a sentence in the proceedings of the Reformed Church in Middelburg of 1607, where it is said that Beeckman "had abstained from listening to God's word in our Dutch church" (hem afgehouden hadden . . . van het gehoor van Godes woort in onse Nederlantsche kercke) (*Journal*, 4:16, n. 5).

19. Ibid., 1:2. On Browne and his followers, see B. R. White, *The English Separatist Tradition from the Marian Martyrs to the Pilgrim Fathers* (London: Oxford University Press, 1971), 44–66; J. G. de Hoop Scheffer, "De Brownisten te Amsterdam gedurende den eersten tijd van hunne vestiging in verband met het ontstaan van de broederschap der Baptisten," in *Verslagen en Mededeelingen der Koninklijke Akademie van Wetenschappen*, Afdeeling Letterkunde, Tweede Reeks, 10 (1881): 203–80, 302–99; F. J. Powicke, *Robert Browne: Pioneer of Modern Congregationalism* (London, 1910); *Nieuw Nederlandsch Biografisch Woordenboek*, vol. 10, cols. 152–55; Ch. Burrage, *The Early English Dissenters in the Light of Recent Research (1550–1641)*, 2 vols. (Cambridge: Cambridge University Press, 1912), 1:108, mentions several gatherings in private houses in Middelburg where local people were also present. See also A. F. Scott Pearson, *Thomas Cartwright and Elizabethan Puritanism, 1535–1603* (Cambridge: Cambridge University Press, 1925), 214.

20. On Veere and its Latin school, see Fortgens, "De Latijnsche scholen te Veere en Arnemuiden," 55–56.

21. For a recent discussion of note taking as a pedagogical device, see Ann M. Blair, *Too Much to Know: Managing Scholarly Information before the Modern Age* (New Haven: Yale University Press, 2010), 62–117. Often the owner of a commonplace book kept a second notebook in which he stored remarks on things he saw, heard, or read about. In Beeckman's case, the commonplace book was used as an "idea book."

22. See the manuscript of Beeckman's notebooks, f. 45r (*Journal*, 1:101). *De Bello Gallico* was then widely read in grammar schools, as it is to this day.

23. Biesius's wife is mentioned as a widow for the first time on March 16, 1607 (*Journal*, 1:199, n. 3), and Beeckman mentions that Biesius died "in initio 18 anni meae aetatis," which means after December 10, 1606.

24. Manuscript of Beeckman's notebooks, f. 296r.

25. *Journal*, 4:16–17.

26. On the university program, see W. Th. M. Frijhoff, *La société néerlandaise et ses gradués, 1575–1814: Une recherche sérielle sur le statut des intellectuels à partir des registres universitaires* (Amsterdam: APA - Holland University Press, 1981), 13–51; Olaf Pedersen, "Tradition and Innovation," in Walter Ruëgg, ed., *A History of the University in Europe*, 4 vols. (Cambridge: Cambridge University Press, 1992–2011), 2:452–70; for Leiden, see Th. H. Lunsingh Scheurleer and G. H. M. Posthumus Meyjes, eds., *Leiden University in the Seventeenth Century* (Leiden: Universitaire Pers Leiden, 1975); Willem Otterspeer, *Groepsportret met Dame, I: Het bolwerk van de vrijheid; De Leidse universiteit, 1575–1672* (Amsterdam: Bert Bakker, 2000), 221–42.

27. On the theological faculty in the early seventeenth century, see Carl O. Bangs, *Arminius: A Study in the Dutch Reformation* (Nashville: Abingdon Press, 1971), 231–355; G. P. van Itterzon, *Franciscus Gomarus* (The Hague: Martinus Nijhoff, 1930), 78–199; Otterspeer, *Bolwerk van de vrijheid*, 243–55. How students experienced the conflicts is revealed in some student notes; see H. W. Tydeman, "Caspar Sibelius, in leven predikant te Deventer, volgens zijne onuitgegeven eigen-levensbeschrijving," *Godgeleerde Bijdragen* 23 (1849): 481–533; A. C. Duker, *Gisbertus Voetius*, 3 vols. (Leiden: Brill, 1910), 1:85–86.

28. *Journal*, 1:217; cf. 1:194. On Van den Broecke, or Vanden Brouck (1566–1626), see Briels, "Zuidnederlandse onderwijskrachten" (1972), 290. In 1609 Van den Broecke published his *Instructie der zee-vaert, door de gheheele werelt. Hier bij is oock een aanhangsel, om de astrolaby catholicum te leeren trecken; voorts, soo komt hier het voornaemste, dat eenen ingenieur ende landt-meter van noode is* (Rotterdam: Abraham Migoen, 1609; 2nd ed. 1610), with a dedication to the city council of Vlissingen. Copies in the university libraries of Groningen and Leiden. At several places Van den Broecke acknowledged the help rendered to him by Willebrord Snellius, "expert in Mathematics" and the son of the Leiden professor of mathematics Rudolph Snellius. So perhaps Willebrord or Rudolph Snellius sent Beeckman to Van den Broecke so he could acquire a decent background in elementary mathematics before embarking on the more sophisticated mathematics Snellius Sr. taught in Leiden.

29. The most famous of all mathematical practitioners was the Amsterdam schoolmaster Willem Bartjens, who is still a household name in Holland and whose skill is recalled in the phrase *rekenen als Bartjens*, which means "to calculate like Bartjens."

30. *Journal*, 4:17–19. Beeckman later did not count Snellius among his teachers; only Van den Broecke was mentioned as such (ibid., 1:217; 2:84). This, however, was true only in a formal sense, because Beeckman did indeed take optional courses with Snellius—for instance, on the optics of Peter Ramus (ibid., 1:6). The reading list Beeckman got from Snellius is undated, but most likely Snellius drew up the list in the fall of 1607. Snellius was a great admirer of Ramus, and as soon as he laid hands on Friedrich Reisner's edition of Ramus's *Optica* in the summer of 1608, he was very enthusiastic about it; see J. A. Vollgraf, "Brieven van Rudolph en Willebrord Snellius," *Leidsch Jaarboekje* 11 (1914): 103–11. In 1609 he devoted one of his courses to the book. On Beeckman's list, however, the book is missing, so the list must date from before the summer of 1608. Given the fact that Beeckman studied the books on the list during a very severe winter and knowing that the winter of 1607-8 was indeed severe, we

may safely conclude that Beeckman was given the list in the fall of 1607. On Rudolph Snellius, see Liesbeth de Wreede, *Willebrord Snellius (1580–1626): A Humanist Reshaping the Mathematical Sciences* (Utrecht: printed by the author, 2007), 36–46.

31. *Journal*, 3:269–70. In comments on the winter of 1607–8 in one of his meteorological notes, Beeckman relates that it was so cold "that people could not remember it had ever been so cold. Peopled walked on ice from Harlingen [in Friesland] to Amsterdam and from Wieringen to Texel. In England people drove coaches with six horses on the Thames and on the 14th, 15th, and 16th of January it was so cold that many people froze to death" (dat by menschen gedinken so een vorst niet geseijn en was men liep op het ijs van Haerlingen tot Amsterdam ende van Wieringhen tot Texel. Men reet in Engeland den Theems meer dan 6 (coetsen) de peerde ende was so koudt op den 14en, 15en ende 16en jan. datter doen veel menschen vervrosen ter doot) (manuscript of Beeckman's notebooks, f. 10r). On the severe winter of 1607–8, see Adriaan M. J. de Kraker, "The Little Ice Age: Harsh Winters between 1550 and 1650," in Pieter Roelofs et al., *Hendrick Avercamp: Master of the Ice Scene* (Amsterdam: Rijksmuseum, 2009), 23–29, esp. 26–28. Avercamp painted his first dated winter scene right after the winter of 1607–8.

32. *Journal*, 3:220, 220–21.

33. Ibid., 1:217–18.

34. Ibid., 1:2. On Ainsworth (1570–1622), who had come to Amsterdam with a congregation of Brownists in 1593 and who was a well-respected Hebraist even in Leiden, see *Nieuw Nederlandsch Biografisch Woordenboek*, vol. 9, cols. 16–18. Beeckman once again visited Amsterdam in February 1610 (*Journal*, 1:2, n. 2).

35. *Journal*, 4:17.

36. H. H. Kuyper, *De opleiding tot den Dienst des Woords bij de Gereformeerden* (The Hague: Martinus Nijhoff, 1891), 512–13; G. Groenhuis, *De predikanten: De sociale positie van de gereformeerde predikanten in de Republiek der Verenigde Nederlanden voor 1700* (Groningen: Wolters-Noordhoff, 1977), 168.

37. H. J. de Jonge, "The Study of the New Testament," in Scheurleer and Posthumus Meyjes, *Leiden University in the Seventeenth Century*, 66–67.

38. *Journal*, 4:20. In the early seventeenth century, it was not uncommon to continue one's studies in this way. See Duker, *Voetius*, 1:78, note; De Jonge, "Study of the New Testament," 67.

39. *Journal*, 4:19; cf. ibid., 1:217. The Beeckman brothers both left Leiden just before the large majority of the theological students submitted a request to the States of the province of Holland and the curators of Leiden University not to nominate the German theologian Conrad Vorstius as Arminius's successor. Vorstius was suspected of being an Anti-Trinitarian (someone who denied the Holy Trinity). Would Isaac and Jacob Beeckman have signed this request, or did they leave Leiden because they did not want to take sides in the conflict? As is clear from later evidence, Isaac Beeckman clearly belonged to the orthodox party. Yet he had his own grudge against the orthodox Calvinists and, above all, disliked getting entangled in theological controversies. Therefore, it would have been quite natural for him to avoid a situation in which he would have been forced to speak out for one of the two parties. It is curious to realize that while Isaac Beeckman left Leiden to return to Middelburg and Jacob Beeckman traveled from Leiden to Franeker, the young David Gorlaeus moved from Franeker to Leiden, where he would write his *Exercitationes* (published in 1620) and his *Idea* (published in 1651), both tracts in which atomistic doctrines are used in defense of Vorstius. Christoph

Lüthy, *David Gorlaeus (1591–1612): An Enigmatic Figure in the History of Philosophy and Science* (Amsterdam: Amsterdam University Press, 2012).

40. *Journal*, 1:218.

41. Ibid., 1:vii; 4:20–21.

42. On this academy, see Pierre-Daniel Bourchenin, *Études sur les académies protestantes en France au XVIe et au XVIIe siècle* (Geneva: Slaktkine Reprints, 1969 [Paris, 1882]), 137–46; L.-J. Méteyer, *L'académie protestante de Saumur* (Paris: La Cause, 1934). In 1612, the professor of theology at Saumur was the Scotsman Robert Boyd of Trochregd, while his countryman Marc Duncan lectured on philosophy. In 1612 Duncan published his *Institutionis logicae libri quinque, in usum Academiae Salmurensis*, a short neo-Aristotelian textbook on logic that in 1625 inspired the Leiden professor of philosophy Franco Petri Burgersdijk when he had to compose a textbook on logic for the Latin schools in Holland. On Duncan, see Joseph Prost, *La philosophie à l'académie protestante de Saumur (1606–1685)* (Paris: Paulin, 1907), 11–45.

43. *Journal*, 1:12–13. The topic of Beeckman and the telescope will be dealt with in chapter 3.

44. *Journal*, 4:28.

45. On the production of candles, see Noel Chomel, *Algemeen huishoudelijk, natuur-, zedekundig- en konst-woordenboek*, 7 vols. (Leiden: Le Maire, 1778–93), 3:1370–73; *Volledige beschrijving van alle konsten, ambachten, handwerken, fabrieken, trafieken, derzelver werkhuizen, gereedschappen*, 24 parts in 9 vols. (Dordrecht: Blussé, 1788–1820), part 3: *De kaarsenmaker* (1789).

46. *Journal*, 1:58, n. 1; 59, n. 3; 62, 173–74.

47. Ibid., 1:218.

48. Ibid., 1:59.

49. These observations were not included in De Waard's edition of the *Journal*: 1:xxxvi.

50. Ibid., 1:xxxvii–xxxviii.

51. Ibid., 2:9.

52. J. J. Westendorp Boerma and C. A. van Swigchem, *Zierikzee vroeger en nu* (Bussum: Fibula-Van Dishoeck, 1972), 57–60. One of the minor poets in Zierikzee was Abraham van der Meer (1582–1632), rector of the Latin school since 1610. See *Nieuw Nederlandsch Biografisch Woordenboek*, vol. 2, cols. 890–91. On his school, see H. W. Fortgens, "De Latijnse scholen te Zierikzee, Brouwershaven en Tolen," *Archief Zeeuwsch Genootschap 1955*, 71–86. Isaac Beeckman remained in contact with Van der Meer after leaving Zierikzee (*Journal*, 1:242). Among the regents of Zierikzee, the interest in natural inquiry was restricted to its edifying potential, as one can see from Adriaen Hoffer's *Nederduytsche poemata* of 1635, the best-known product of the literary life in Zierikzee. See Meertens, *Letterkundig leven*, 326–34.

53. On the contribution of Zeeland to Pietism, see ibid., 168–85.

54. Ibid., 173–78; W. J. op 't Hof, *Willem Teellinck (1579–1629): Leven, geschriften en invloed* (Kampen: De Groot Goudriaan, 2008).

55. P. J. Meertens, "Godefridus Cornelisz. Udemans," *Nederlandsch Archief voor Kerkgeschiedenis*, Nieuwe Serie, 28 (1935): 65–106; E. Beins, "Die Wirtschaftsethik der calvinistischen Kirche der Niederlanden 1565–1650," *Nederlandsch Archief voor Kerkgeschiedenis*, Nieuwe Serie, 24 (1931): 81–156.

56. N. van der Blom, *Grepen uit de geschiedenis van het Erasmiaans Gymnasium 1328–1978* (Rotterdam: W. Backhuys, 1978), 63–64, 70–74.

57. *Journal*, 2:301; 4:79–80, 145. According to the inventory of his library, Isaac Beeckman owned several copies of the works of Teellinck but none of Udemans.

58. Jacob Beeckman was no doubt appointed on the recommendation of Abraham van der Meer, who had been rector of the Latin school of Veere from 1607 to 1610 (as the successor of Biesius). Jacob Beeckman was appointed in April but only moved to Veere in July or August (*Journal*, 4:37).

59. Ibid., 1:217; 4:35.

60. Ibid., 1:219.

61. Ibid., 1:173.

62. Ibid., 1:190.

63. Ibid., 1:126.

64. Ibid., 1:134.

65. As explained in the introduction, De Waard left out these genealogical notes from his edition of the *Journal* because they were—to his mind—of no scientific interest. However, he indicated where the reader could find them: in the manuscript of Beeckman's notebooks, ff. 47v–50r, 234v–315v.

66. Elisabeth Moreau, "Le substrat galénique des idées médicales d'Isaac Beeckman (1616–1627)," *Studium* 4 (2011): 137–51.

67. Charles Scott Sherrington, *The Endeavour of Jean Fernel, with a List of the Editions of His Writings* (Cambridge: Cambridge University Press, 1946), 64–65. Fernel was professor of medicine at Paris from 1534 until his death. He was also physician to the king of France. See also K. E. Rothschuh, "Technomorphes Lebensmodell contra Virtus-Model (Descartes gegen Fernel)," *Sudhoffs Archiv* 54 (1970): 337–54.

68. *Journal*, 1:218 (on Fernel); ibid., 1:112, 128–29.

69. Ibid., 2:84.

70. Students who went abroad to get a doctoral degree usually traveled in small groups (for safety reasons). Beeckman's companions were his uncle Jan Pietersz. van Rhee (his mother's brother, owner of an oil mill, who settled in Amsterdam in late 1618), Hendrick Somer (son of a burgomaster of Veere, former pupil of Jacob Beeckman, soon to be a student of medicine at Leiden University, but finally becoming a merchant in Middelburg), and Caspar Zegers (from The Hague). On the visits of Dutch students to foreign universities, see Frijhoff, *La société néerlandaise*, 83–94. And on Caen in particular, see ibid., 86, 136. Frijhoff claims that between 1610 and 1620 no less than twenty-two Dutch students took a medical degree at Caen. See also W. Th. M. Frijhoff, "Nederlandse promoties in de geneeskunde aan Franse universiteiten (zestiende-achttiende eeuw)," *Jaarboek van het Centraal Bureau voor Genealogie* 60 (2006): 75–111. According to Paul Dibon, *Le voyage en France des étudiants néerlandais au XVIIème siècle* (The Hague: Martinus Nijhoff, 1963), 24, of the 255 physicians of the Amsterdam Collegium Medicum in the years 1628–76, at least 42 earned their degree in France, of whom 19 (almost 50%) had done so at Caen.

71. *Journal*, 4:40–42.

72. The three remaining pages are included in De Waard's edition of the *Journal*, 4:42–44. Beeckman dedicated his dissertation to his father, his close friend and brother-in-law Jacob Schouten, and his brother Jacob, who is called "frater germanus et Pylades suus" (Pylades was the close friend and helper of Orestes, the son of Agamemnon and Clytemnestra; in classical times, the two were regarded as exemplary friends). The only surviving copy is in the British Museum. For the possible provenance of this copy, see chapter 3, note 71. "Malaria tertiana," the topic of Beeckman's dissertation, was of special interest for someone from Zeeland, where the disease was endemic. Priester, *Geschiedenis van de Zeeuwse landbouw*, 91.

73. *Journal*, 1:205.

74. In his notebooks, Beeckman gives a slightly different set of corollaries (ibid., 1:200–201). Instead of a corollary on the motion of a projectile, Beeckman put forward a logical proposition: "Homo aut canis non est infima species logica" (Man and dog are not the lowest logical entities), which is a strong indication that Beeckman was much more preoccupied at this stage with logical analysis than with empirical research. The proposition on the possibility of squaring the circle is also missing from the list of corollaries in the notebooks. Beeckman must have drafted this list before the dissertation defense, and apparently changed his mind shortly before sending the dissertation with the corollaries to the printer.

75. Beeckman's dissertation was the first *publication* in which the basic idea of inertia in the modern sense was formulated (the idea that not movement as such but only a change in movement needs an explanation). The idea itself however had been formulated in the notebooks much earlier, as was the case with the other corollaries.

76. That the faculty was impressed with Beeckman's talents was communicated to me by Willem Frijhoff, who studied the register extensively. He also mentioned that from the way De Waard refers to the page numbers and quotes from the Matrologium, it is clear that he has not seen the register himself. The correct text is reproduced in Henri Prentout, "Esquisse d'une histoire de l'université de Caen," in Henri Prentout et al., *1432–1932: L'université de Caen; Son passé, son présent* (Caen: Malherbe de Caen, 1932), 131. The deviations are not essential, however, the most important being De Waard's substitution of "praestantissimus" (most excellent) by "peritissimus" (most experienced).

77. *Journal*, 4:49.

78. Ibid., 1:223; 4:48.

79. Ibid., 1:225, 228.

80. Pieter Cools had married Elisabeth Pieters van Rhee, a sister of Beeckman's mother, in 1595. At that time he was called "jonckgeselle van Blanckeberch," that is, "a young man from Blanckenberch," a small town at the coast north of Bruges. In 1597 he is mentioned as a "tanner" (huyvetter) in Middelburg. Notwithstanding De Waard's statement to the contrary (*Journal*, 1:228, n. 3), Cools's name appears in the archives at Breda. He became a "burgher" (poorter) of Breda on July 22, 1613, Municipal Archives, Breda, Poorterboek 1534–1632, inv. no. 2111 (cf. *Journal*, 4:7).

81. *Journal*, 1:228.

82. Though biased, Adrien Baillet, *La vie de monsieur Des-Cartes*, 2 vols. (Paris: Daniel Horthemels, 1691; reprint, Paris: La Table Ronde, 1992; trans. as *The Life of Monsieur Des Cartes* [London: R. Simpson, 1693]), remains an essential source for Descartes's early life. For modern biographies, see the bibliographical essay at the end of this volume.

83. Every major biography of Descartes deals extensively with this formative period of Descartes's life and therefore with his first meeting with Beeckman. Most illuminating are Stephen Gaukroger, *Descartes: An Intellectual Biography* (New York: Oxford University Press, 1995), 68–103; Richard A. Watson, *Cogito Ergo Sum: The Life of René Descartes* (Boston: David R. Godine, 2002), 78–92; Desmond M. Clarke, *Descartes: A Biography* (Cambridge: Cambridge University Press, 2006), 41–52.

84. Daniel Lipstorp, *Specimina philosophiae cartesianae* (Leiden: Elsevier, 1653), 76–78, as quoted in *Oeuvres de Descartes*, publiées par Charles Adam et Paul Tannery, 13 vols. (Paris: Vrin, 1964–75 [Paris: Léopold Cerf, 1897–1957]), 10: 47–48. Cf. Baillet, *La vie de monsieur Des-Cartes*, 24–28. The source for this story was probably the Leiden mathematician Frans van Schooten

Jr., who drew the figures for Descartes's *Principia philosophiae* (1644) and translated Descartes's *Géométrie* in Latin. For an interesting commentary, see John R. Cole, *The Olympian Dreams and Youthful Rebellion of René Descartes* (Urbana: University of Illinois Press, 1992), 81.

85. *Journal*, 4:201-2.

86. Ibid., 1:237. See Clarke, *Descartes*, 42-43.

87. *Journal*, 1:237, 242, 244, 247, 257, 258.

88. J. Sirven, *Les années d'apprentissage de Descartes* (1596-1628) (Albi: Imprimerie coopérative du Sud-Ouest, 1928), 57.

89. *Journal*, 1:257.

90. Ibid., 1:244: "He said he had never found anyone, except me, who was used to study in this way, which I find very delightful, and who combines in an exact way mathematics and physics" (Dicit se nunquam neminem reperisse, praeter me, qui hoc modo, quo ego gaudeo, studendi utatur accurateque cum Mathematica Physicam jungat).

91. Ibid., 1:257.

92. Ibid., 1:296. Beeckman first mentions the receipt of the *Compendium* in a note dated January 2 at Geertruidenberg; much later in a letter to Mersenne, he refers to it as a book that was sent to him: "D. Des Chartes, amicus noster, in libello suo, quem de Musica conscriptum ad me *misit*. . . ." (Beeckman to Mersenne, October 1, 1629, ibid., 4:162).

93. Ibid., 4:56.

94. Ibid.

95. Ibid., 1:xxxviii; 4:56. Much later, Constantijn Huygens also owned a copy, which included Descartes's dedication to Beeckman (University Library Leiden, Ms. Hug. 29.a). In a letter to Descartes of September 8, 1637, Huygens states that back then he had been jealous of this man, "for whom you once wrote the Treatise on Music" (qu'il y a longtemps que je suis jaloux de cest honeste homme, en faveur duquel vous avez autrefois escrit le Traicté de la musique) (*Oeuvres de Descartes*, 1:396). In 1650, the *Compendium* was published as the *Musicae compendium*.

96. *Journal*, 4:56, 60; 1:281-82.

97. Ibid., 4:63-64.

98. Ibid., 4:64-66.

99. Ibid., 4:62 (translation by John R. Cole).

100. For a psychologically sophisticated reading of the Descartes-Beeckman correspondence, see Cole, *Olympian Dreams*, 114-28.

101. Watson, *Cogito, ergo sum*, 89-90.

102. Clarke, *Descartes*, 48.

103. Beeckman to Descartes, May 6, 1619, *Journal*, 4:65.

104. Ibid., 4:65 (translation by John R. Cole).

105. Cole, *Olympian Dreams*, 125-26.

CHAPTER 2: SCHOOLTEACHER AND CRAFTSMAN, 1619-1627

1. Jonathan I. Israel, *The Dutch Republic: Its Rise, Greatness, and Fall, 1477-1806* (Oxford: Clarendon Press, 1995), 460-65.

2. *Journal*, 4:66. The events in Utrecht have been described by Izaäk Vijlbrief, *Van antiaristocratie tot democratie. Een bijdrage tot de politieke en sociale geschiedenis der stad Utrecht* (Amsterdam: Querido, 1950). See also Benjamin Jacob Kaplan, *Calvinists and Libertines: Confession and Community in Utrecht, 1578-1620* (Oxford: Clarendon Press, 1995). Aemilius (1589-1660) had been principal of the Latin school in Dordrecht from 1615 to 1619. With one

short interruption, he would stay at the Latin school and the University of Utrecht until his death. See *Nieuw Nederlandsch Biografisch Woordenboek*, vol. 1, cols. 38–39.

3. *Journal*, 1:xiii, n. 3.

4. Ibid., 4:68.

5. Ibid., 2:1; 4:67.

6. Ibid., 2:4.

7. For further information on the Latin school in Utrecht, see A. Ekker and C. D. J. Brandt, *De Hieronymusschool te Utrecht*, 2 vols. (Utrecht: Bosch, 1863–1924); G. W. Kernkamp, *De Utrechtse Academie, 1636–1815* (Utrecht: Oosthoek, 1936), 25–40; E. P. de Booy, *Kweekhoven der Wijsheid. Basis- en vervolgonderwijs in de steden van de provincie Utrecht van 1580 tot het begin der 19e eeuw* (Zutphen: De Walburg Pers, 1980), 121–27. For Latin schools in general see also H. W. Fortgens, *Schola Latina: Uit het verleden van ons voorbereidend hoger onderwijs* (Zwolle: Tjeenk Willink, 1958), 9–45.

8. *Journal*, 3:221.

9. Ibid., 2:5, 19.

10. Ibid., 2:84.

11. Ibid., 2:25, 44, 122.

12. Ibid., 2:36–37.

13. Previously Nota had obtained patents from the States General for a new method to repair broken canons and kettles (March 6, 1617) and for a new kind of furnace, plus a way to clean streets and to ventilate houses (February 16, 1618). G. Doorman, ed., *Octrooien voor uitvindingen in de Nederlanden uit de 16e–18e eeuw* (The Hague: Nijhoff, 1940), 134, 137–38. See also Karel Davids, *The Rise and Decline of Dutch Technological Leadership: Technology, Economy, and Culture in the Netherlands*, 2 vols. (Leiden: Brill, 2008), 1:142.

14. *Journal*, 2:38–40.

15. J. B. Kan, *Geschiedenis van het Erasmiaansch Gymnasium* (Rotterdam: Nijgh en Van Ditmar, 1884), 23–24; *Nationaal Biografisch Woordenboek* (of Belgium), vol. 6, cols. 76–79. At first, the Rotterdam regents had settled on Adriaan Smout (1580–1646), a militant Counter-Remonstrant minister, but Smout preferred to accept a post in Amsterdam. The fact that the regents wanted him to become the new principal, testifies to their preference for someone with undisputed orthodox opinions. Apparently, Jacob Beeckman fulfilled this condition too. On Smout, see *Nieuw Nederlandsch Biografisch Woordenboek*, vol. 10, cols. 941–44.

16. *Journal*, 2:152–53, 156; 4:70–73.

17. Ibid., 2:204. The number of boarding pupils in Utrecht had been higher, according to De Booy, *Kweekhoven der Wijsheid*, 122–23, somewhere between seventy-five and one hundred. In the 1630s, in Dordrecht Beeckman would have about sixty boarding pupils (*Journal*, 3:369).

18. P. C. Molhuysen, ed., *Bronnen tot de geschiedenis der Leidsche universiteit*, 7 vols. (The Hague: Martinus Nijhoff, 1913–24), 3:230*.

19. Ent would also visit Beeckman after his move to Dordrecht (*Journal*, 3:24).

20. The fact that Hortensius went to school at Rotterdam with Beeckman is mentioned by Reinier Boitet, *Beschrijving der stadt Delft* (Delft: Reinier Boitet, 1729), 726.

21. *Journal*, 2:i–xxi, 291–92. Cf. C. M. J. van den Heuvel, *"The Huysbou": A Reconstruction of an Unfinished Treatise on Architecture, Town Planning and Civil Engineering by Simon Stevin* (Amsterdam: Koninklijke Nederlandse Akademie van Wetenschappen, 2005), esp. 166.

22. Municipal Archives Rotterdam, Resolutions of the City Council, July 17, 1621. See also Kan, *Erasmiaansch Gymnasium*, 26–28.

23. *Journal*, 2:311.

24. Molhuysen, *Bronnen*, 3:230*.

25. Ibid. According to the Leiden rector and senate, Jacob and Isaac Beeckman were on a par with Ubbo Emmius in Groningen (headmaster from 1594 to 1613 and afterward professor of Greek and history and first rector magnificus of the newly founded University of Groningen) and Jodocus Honingius at Harderwijk (rector of the Latin school from 1603 to 1637 and as such also rector of the illustrious school).

26. Municipal Archives Rotterdam, Resolutions of the City Council, July 15, 1624, September 7, 1624, September 8, 1624. On the new program, see E. J. Kuiper, *De Hollandse "Schoolordre" van 1625: Een studie over het onderwijs op de Latijnse scholen in Nederland in de 17e en 18e eeuw* (Groningen: Wolters, 1958).

27. *Journal*, 2:373 (this note dates from the fall of 1626).

28. Ibid.

29. Ibid., 2:249.

30. Ibid., 2:264, 303. Weymans originated from Sint-Truiden, between Brussels and Maastricht.

31. Ibid., 2:173, 179. Beeckman most likely agreed to repair the water pipes in the brewery in Veere out of friendship. The brewery was located in the same street as the Latin school, the Wagenaar Street, and Beeckman probably knew the owner from the time that he visited his brother Jacob when he was principal of the Latin school.

32. Ibid., 2:122. See also the remarks on a magnifying glass (ibid., 2:298) and a "perpetual fountain" (ibid., 2:208).

33. C. te Lintum, "De oprichting van de Rotterdamsche kamer der West-Indische Compagnie," *Rotterdamsch Jaarboekje 1910*, 104.

34. For the general history of Rotterdam, see H. C. Hazewinkel, *Geschiedenis van Rotterdam* (Amsterdam: Joost van den Vondel, 1940–42); Arie van der Schoor, *Stad in aanwas. Geschiedenis van Rotterdam tot 1813* (Zwolle: Waanders, 1999). R. Bijlsma, *Rotterdams welvaren 1550–1650* (The Hague: Nijhoff, 1918) is also still useful. In 1622 Rotterdam officially counted 19,532 inhabitants, only Amsterdam, Leiden, Haarlem, and Delft being more populous (in the province of Holland). Therefore, Beeckman was not far off the mark when in 1626 he rounded off the number of residents at 20,000 (*Journal*, 2:356). See J. G. van Dillen, "Summiere staat van de in 1622 in de provincie Holland gehouden volkstelling," *Economisch-historisch jaarboek* 21 (1940): 167–89.

35. J. H. Kernkamp, *Johan van der Veeken en zijn tijd* (The Hague: Martinus Nijhoff, 1952); R. Bijlsma, "De Zuidnederlandsche immigranten en de textiel-industrie in oud-Rotterdam," *Vragen van den Dag* 31 (1916): 858–67; Bijlsma, "De lakencompagnie der Van Berckels," *Rotterdamsch Jaarboekje 1913*, 82–88; Bijlsma, *Rotterdams welvaren*, 101–6.

36. As quoted by A. Th. van Deursen, *Het kopergeld van de Gouden Eeuw* (Assen: Van Gorcum, 1978–1980), 1:21.

37. For a general survey of technological development in the Dutch Republic, see Davids, *Technological Leadership*, 57–201.

38. For the affair, see *Journal*, 2:350–59. The affair is also mentioned by Doorman, *Octrooien*, 166. See also G. Doorman, *Techniek en octrooiwezen in hun aanvang: Geschiedkundige aanvullingen* (The Hague: Martinus Nijhoff, 1953), 60. Beeckman did not learn how much Puyck was willing to pay until long after the affair; see *Journal*, 3:13–14.

39. Ibid., 2:358.

40. Ibid., 2:352.

41. Ibid., 2:358.

42. Ibid., 2:353, 355–56.

43. Ibid., 2:359. Although he suspected something was amiss, Beeckman, of course, did not know in 1626 that the spire would become so rotten that it had to be torn down immediately in 1642. See Hazewinkel, *Geschiedenis van Rotterdam*, 3:223.

44. *Journal*, 2:359. In April 1625, during the siege, the defenders of Breda built a dam in a nearby river to arrest the flow of water and to flood the Spanish quarters. The dam, however, proved to be inadequate and was broken by the enormous flow of water, as a result of which the defenders' quarters, not those of the Spanish, were flooded. After a while, the Dutch forces tried it again, with the same disastrous results. The surrender of the city was then only a matter of time, and on June 2, 1625, the Spanish general Spinola took over the city. See Th. E. van Goor, *Beschryving der stadt en lande van Breda* (The Hague: Jacobus vanden Kieboom, 1744), 168. Beeckman's remarks date from August 1625.

45. For the minutes of the Collegium Mechanicum, see *Journal*, 2:429–56. Cf. Klaas van Berkel, "Het Collegium mechanicum van Isaac Beeckman," *Spiegel Historiael* 15 (1980): 336–41.

46. Beeckman's Collegium Mechanicum does however resemble proposals made during the English Civil War by Puritans who wanted to systematize and institutionalize technological improvements for the benefit of the community. An example is the Gymnasium Mechanicum that William Petty wanted to start in 1648. In Petty's Gymnasium, all kinds of skilled practitioners would come together to make an inventory of the technology of their trades and would devise improvements on the basis of sound practices. The plan had a more systematic character than Beeckman's, but the similarities are striking. On Petty's Gymnasium, see Paolo Rossi, *Philosophy, Technology and the Arts in the Early Modern Era* (New York: Harper & Row, 1970), 123; Charles Webster, *The Great Instauration: Science, Medicine, and Reform 1626–1660* (New York: Holmes and Meier Publishers, 1975; 2nd ed., Oxford: Lang, 2002), 363.

47. M. J. van Lieburg, "Johannes Furnerius (1582–1668), een musicerende scheepschirurgijn, en zijn carrière tot stads- en landsdokter te Rotterdam," *Rotterdams Jaarboekje 1978*, 223–42. Possibly "Mr. Pieter" is identical with Hugh Peters, the English preacher, a personal friend of Beeckman (see note 84).

48. *Nieuw Nederlandsch Biografisch Woordenboek*, vol. 2, cols. 1356–58. In the literature, Stampioen is frequently mixed up with his son, who has the same name but is usually called "the younger" (De Jonge). See, for instance, *Dictionary of Scientific Biography*, 12:610–11. The confusion has been cleared up by J. Mac Lean, "Stampioen," *De Nederlandsche Leeuw* 74 (1957): cols. 323–28. On Stampioen Jr., see *Nieuw Nederlandsch Biografisch Woordenboek*, vol. 2, cols. 1358–60. Stampioen Jr., who was to become a well-known mathematician, knew Greek and Latin, which he probably learned from Jacob and Isaac Beeckman. Perhaps Beeckman got to know the elder Stampioen through his son. The boy was born in 1610 and turned sixteen in 1626. Cf., however, C. A. Davids, *Zeewezen en wetenschap. De wetenschap en de ontwikkeling van de navigatietechniek in Nederland tussen 1585 en 1815* (Amsterdam-Dieren: De Bataafsche Leeuw, 1986), 457, n. 18.

49. In 1644 Stampioen's second marriage was not consecrated in the Reformed Church but concluded only before the Rotterdam magistrates. After his death in 1660, one of his sons was not endowed with more than the minimum share of the inheritance because he had

accused his father of heresy. Among the executors of his will was Abraham Beeckman Jr., Isaac's younger brother, who at that time was principal of the Latin school in Rotterdam. Mac Lean, "Stampioen," col. 325. The fact that the elder Stampioen belonged to the Remonstrant Church once again draws our attention to the fact that numerous mathematical practitioners in the Dutch Republic were Mennonites, Remonstrants, or members of other dissident denominations. Cf. H. A. M. Snelders, "Alkmaarse natuurwetenschappers uit de 17e en 18e eeuw," in *Van Spaans beleg tot Franse tijd. Alkmaars stedelijk leven in de 17e en 18e eeuw* (Zutphen: De Walburg Pers, 1980), 101–22.

50. *Journal*, 2:349–50, 359–60.

51. Ibid., 2:440.

52. Ibid., 2:430.

53. Ibid., 2:431–32.

54. Ibid., 2:432. There is no archival evidence relating to this mill. Perhaps the merry-go-round that Beeckman referred to never got further than the experimental stage.

55. Ibid., 2:446.

56. Ibid., 2:445–46.

57. Ibid., 2:432–34, 444. The sails on the wings of the merry-go-round near Rijswijk were adjustable. On running backwards, the sails would collapse, so as not to cause resistance.

58. Ibid., 2:447.

59. Ibid. In Dutch, "popmeulenties."

60. Ibid., 3:90–91.

61. Ibid., 2:439.

62. Especially for Stevin's remarks on water scouring (Waterschueringh), see *Journal*, 2:417–18. See E. J. Dijksterhuis, *Simon Stevin: Science in the Netherlands around 1600* (The Hague: Martinus Nijhoff, 1970), 100–101.

63. *Journal*, 2:439–40.

64. Ibid., 2:440–41.

65. Ibid., 2:442. The members of the Collegium on the other hand were aware of the adverse effects that drastic measures might have. Too many dams and jetties might narrow the bed of the river too much, increasing pressure on the dikes (and possible breaches) in case of high water or excessive rainfall (ibid., 2:442–43).

66. Ibid., 2:440.

67. Ibid., 2:444.

68. Ibid., 2:447–48.

69. Municipal Archives Rotterdam, Resolutions of the City Council, January 4, 1627. The dam can be seen on the map of the High Water Board Schieland made by Stampioen Jr. in 1660. Municipal Archives Rotterdam, Historisch-Topografische Atlas, no. 610.13.492.61 (see Van Berkel, "Collegium mechanicum," 341).

70. *Journal*, 2:448.

71. Ibid., 2:442.

72. Ibid., 2:440.

73. Ibid., 2:454.

74. Ibid., 2:454–55.

75. Ibid., 2:328; 4:85–86. The school in Den Briel was much smaller than the school in Rotterdam: it had only thirty pupils.

76. Ibid., 3:6.

77. A. Th. van Deursen, *Bavianen en Slijkgeuzen. Kerk en kerkvolk ten tijde van Maurits en Oldenbarnevelt* (Assen: Van Gorcum, 1974), 346–71.

78. Ibid., 134, 309, 371. On the history of the persecution of the Remonstrants in Rotterdam, see Hazewinkel, *Geschiedenis van Rotterdam*, 3:171–200; Tj. R. Barnard and E. H. Cossee, *Arminianen in de Maasstad. 375 jaar Remonstrantse Gemeente in Rotterdam* (Amsterdam: De Bataafsche Leeuw, 2007), 17–31.

79. Van Leeuwen was a preacher in Rotterdam since 1621. Municipal Archives Rotterdam, Resolutions of the City Council, November 30, 1621. For the history of the erection of a statue for Erasmus in 1622, see N. van der Blom, *Erasmus en Rotterdam* (Rotterdam: Nijgh & Van Ditmar, 1969), 41–49, reprinted in *Florislegium. Bloemlezing uit de Erasmiaanse, Rotterdamse en andere opstellen van drs. N. van der Blom* (Leiden: Brill/Backhuys, 1982), 29–54. See also the resolution of the city council, dated May 25, 1622, as printed in *Rotterdams Jaarboekje 1922*, 26.

80. *Journal*, 2:329, 264, n. 1. All extant documents pertaining to this affair have been reproduced in Beeckman's *Journal* (4:86–114). The proceedings of the Rotterdam church council have not been preserved.

81. After his ordination as a preacher in Amsterdam, on June 29, 1625, Hanecop indeed showed himself to be very lenient toward the Remonstrants, which caused him a lot of trouble with his more militant colleagues (*Nieuw Nederlandsch Biografisch Woordenboek*, vol. 6, cols. 695–96).

82. On Van Goch, see E. A. Engelbrecht, *De vroedschap van Rotterdam 1572–1795* (Rotterdam: Gemeentelijke Archiefdienst, 1973), 134–35.

83. *Journal*, 4:102–3.

84. Probably Beeckman visited the English church during the last years of his stay in Rotterdam (ibid., 2:333). After his departure for Dordrecht, he remained in contact with the Puritans in Rotterdam, and their preacher, Hugh Peters, was a personal friend (ibid., 3:324–42).

85. Ibid., 4:105.

86. Ibid., 2:327, 345.

87. Ibid., 2:341; 4:95.

88. On January 30, 1626, almost two months after the death of Abraham Beeckman Sr., four houses neighboring the parental home "The Two Roosters" were sold, leaving only this house and the adjacent, smaller house called "The Two Little Roosters" in the possession of the family (*Journal*, 4:96). Beeckman's sister Esther, who was married to the candlemaker Louis Vergrue, moved into the former, and Beeckman's mother into the latter house (ibid., 4:108). However, Beeckman's mother died in the house of her son Jacob in Rotterdam on June 25, 1629. She also was buried in Rotterdam (ibid., 3:122; 4:149). The family sold "The Two Little Roosters" house a year after her death.

89. Ibid., 4:77. For more information, see N. van der Blom, *Grepen uit de geschiedenis van het Erasmiaans Gymnasium 1328–1978* (Rotterdam: Backhuys, 1978), 79–80.

90. *Journal*, 4:104; Resolution of the Rotterdam City Council, quoted in Hazewinkel, *Geschiedenis van Rotterdam*, 3:198–99.

91. *Journal*, 3:450. Van Berckel died of tuberculosis, which a few years earlier had also killed Isaac's brother Jacob. Beeckman inserted a report on Van Berckel's case history in the *Journal* (3:446–50). He called him "a man of good and tender conscience," which I take to be a reference to his being a sympathizer of Teellinck, just like Beeckman. Van Berckel seems to have corresponded with the pietist preacher. See W. J. op 't Hof et al., *Eeuwout Teellinck in handschriften* (Kampen: De Groot Goudriaan, 1989), 34–38, 47–51.

92. *Journal*, 4:112–14.

93. Ibid., 3:6.

CHAPTER 3: AMONG PATRICIANS AND PHILOSOPHERS, 1627–1637

1. For a general survey of the history of Dordrecht, see W. Th. M. Frijhoff, ed., *Geschiedenis van Dordrecht*, 3 vols. (Hilversum: Verloren, 1996–2000), esp. vol. 3: *Geschiedenis van Dordrecht van 1572 tot 1813* (1998).

2. P. W. Klein, *Kapitaal en stagnatie tijdens het Hollandse vroegkapitalisme* (Rotterdam: Universitaire Pers, 1967), 6; Peter Burke, *Venice and Amsterdam: A Study of Seventeenth-Century Élites* (London: Temple Smith, 1974), 60–61, 108–12.

3. D. J. Roorda, "The Ruling Classes in Holland in the Seventeenth Century," in J. S. Bromley and E. H. Kossmann, eds., *Papers Delivered to the Anglo-Dutch Historical Conference, 1962* (Groningen: Wolters, 1964), 109–32; H. van Dijk and D. J. Roorda, "Sociale mobiliteit onder regenten in de Republiek," *Tijdschrift voor Geschiedenis* 84 (1971): 306–28; Hugo Soly, "Het "verraad" der 16e-eeuwse burgerij: een mythe?" *Tijdschrift voor Geschiedenis* 86 (1973): 262–80. On De Witt and his family, see H. H. Rowen, *John de Witt, Grand Pensionary of Holland (1625–1672)* (Princeton: Princeton University Press, 1978), revised as *John de Witt, Statesman of the "True Freedom"* (Cambridge: Cambridge University Press, 1986). John de Witt, pensionary of Holland and a well-known mathematician, was Jacob de Witt's son.

4. N. Japikse, "De Dordtse regeringsoligarchie in het midden der 17e eeuw," *Bijdragen voor Vaderlandsche Geschiedenis en Oudheidkunde*, Zesde Serie, 1 (1924): 6–22, esp. 6–14; J. L. van Dalen, *Geschiedenis van Dordrecht*, 2 vols. (Dordrecht, 1931–33), 2:1127; Ivo Schöffer, "Viel onze Gouden Eeuw in een tijdvak van crisis?" *Bijdragen en Mededelingen van het Historisch Genootschap* 78 (1964): 45–74.

5. A. Th. van Deursen, *Het kopergeld van de Gouden Eeuw*, 4 vols. (Assen: Van Gorcum, 1978–80), 1:67.

6. L. van Bos, *Dordrechtsche Arcadia* (Dordrecht: A. Andriesz., 1662), 577.

7. *Journal*, 4:118, n. 4.

8. Ibid., 4:118–21.

9. Ibid., 3:369. G. D. J. Schotel, *De Illustere school te Dordrecht: Eene bijdrage tot de geschiedenis van het schoolwezen in ons vaderland* (Utrecht: Kemink, 1857), 64–65, gives a beautiful description of the school, but because there is no accurate picture of the school, it is difficult to establish whether Schotel's nineteenth-century description is accurate. The pictures of the building in the city maps of Jacob van Deventer (1545), Francesco Guicciardini (around 1590), and Joan Blaeu (between 1640 and 1647) are all different, and none of them includes the tower mentioned by Schotel. Presumably, the former Augustine monastery in which the school was housed was an inconspicuous building.

10. *Journal*, 3:6.

11. For more details, see Klaas van Berkel, *Isaac Beeckman (1588–1637) en de mechanisering van het wereldbeeld* (Amsterdam: Rodopi, 1983), 101, n. 16. The vice-principal earned considerably less, 380 guilders, than the principal.

12. *Journal*, 3:74: "molestissimum et ad omnes meditationes ineptissimum munus." See also ibid., 4:180.

13. Ibid., 4:212, 238.

14. Ibid., 3:32.

15. Ibid., 1:xxiii.

16. Ibid., 4:169. Abraham Beeckman (born on January 15, 1607) had passed through the Latin schools in Veere and Rotterdam while his brother Jacob was principal there. Nothing is known of a university education, and the story that he acquired a degree in law at the University of Paris in 1630 (ibid., 4:238) is highly unlikely because only canonical law was taught in Paris, which was useless for a Protestant. See W. Th. M. Frijhoff, *La société néerlandaise et ses gradués, 1575–1814: Une recherche sérielle sur le statut des intellectuels à partir des registres universitaires* (Amsterdam: APA - Holland University Press, 1981), 86.

17. *Journal*, 3:5.

18. Ibid., 4:122–26.

19. Barlaeus's Amsterdam oration is often misinterpreted. See Klaas van Berkel, "Rediscovering Clusius: How Dutch Commerce Contributed to the Emergence of Modern Science," *Bijdragen en Mededelingen betreffende de Geschiedenis der Nederlanden / Low Countries Historical Review* 123 (2008): 227–36.

20. What the observatory really looked like is hard to tell. Sometimes Beeckman talks about a simple "platform" (myn pladt) (*Journal*, 3:112), at other times he refers to his "tower" (turrus) (ibid., 3:85, 215). The resolutions of the city council do not mention the tower's construction, despite Beeckman's claim that it involved "great expense" (ibid., 3:112).

21. Ibid., 3:85.

22. Ibid.

23. Ibid., 3:153. The purpose of this observation was to determine the exact length (duration) of the eclipse.

24. Ibid., 3:215, 396. When Beeckman referred to a telescope, he used a variety of words. In 1612, during his stay at Saumur, he talked about an "instrument with which one can see small things from afar, in French *lunette* (instrumentum quo e longinquo res parvae videntur et gallice lunette vocatur) (ibid., 1:12). In 1618, this time at Caen, he consulted Hieronymus Sirtorius's *Telescopium* and commented on the use of the diaphragm of what is called an "eye tube" (tubus ocularis) (ibid., 1:208–09), a term he also used in 1622, 1623, and 1626 (ibid., 2:209–10, 247, 346, 396). In Dutch, Beeckman referred to the "verrekycker" or "verresiender" (the instrument to see far) (ibid., 2:294–96); cf. ibid., 2:357, 367; 3:46, 69. Only in the spring of 1629 did the word "telescope" appear in the notebooks (ibid., 3:121). Beeckman also uses this technical term in marginal headings, which he added to his notes around this time.

25. Ibid., 2:186–87.

26. Ibid., 2:361, 366; 3:85.

27. Of course, it would be misleading to speak of a thermometer, because in Drebbel's instrument the influence of the air pressure was not eliminated. See F. Sherwood Taylor, "The Origin of the Thermometer," *Annals of Science* 4 (1942): 129–56; M. K. Barnett, "The Development of Thermometry and the Temperature Concept," *Osiris* 12 (1956): 269–341, esp. 269–89.

28. *Journal*, 1:346. On Drebbel, see F. M. Jaeger, *Cornelis Drebbel en zijne tijdgenooten* (Groningen: Noordhoff, 1922); Gerrit Tierie, *Cornelis Drebbel (1572–1633)* (Amsterdam: Paris, 1932); Vera Keller, "Drebbel's Living Instruments, Hartmann's Microcosm, and Libavius's Thelesmos: Epistemic Machines before Descartes," *History of Science* 48 (2010): 39–74; Keller, "How to Become a Seventeenth-Century Natural Philosopher: The Case of Cornelis Drebbel (1572–1633)," in Sven Dupré and Christoph Lüthy, eds., *Silent Messengers: The Circulation of Material Objects of Knowledge in the Early Modern Low Countries* (Berlin: LIT Verlag, 2011), 125–51. Beeckman did not know Drebbel personally, but his father may very well have. It is reported that in 1601 Drebbel was engaged in the construction of a fountain in a garden

outside one of the Middelburg city gates (Tierie, *Drebbel*, 4). Possibly Drebbel was employed only to design the fountain and hired Beeckman Sr., to do the actual construction. Beeckman also spoke regularly with others who claimed to have personal information concerning Drebbel and his inventions. These people included the philosopher Henricus Reneri (the private tutor of the children of the pensionary of Amsterdam, Adriaan Pauw) (*Journal*, 2:371–72), Beeckman's brother-in-law Justinus van Assche (ibid., 3:3, 302, 358), and Drebbel's son-in-law Johan Sibertus Kuffler (ibid., 3:367).

29. *Journal*, 2:25; 3:439–42.

30. Ibid., 2:202–3, 344, 363, 370–72; 3:203–204, 302–4, 358, 367.

31. Keller, "Drebbel's Living Instruments," 39–40.

32. *Journal*, 3:203–4, 302–3.

33. It is interesting to note that Beeckman never called Drebbel an impostor, whereas others who made equally extravagant claims were denounced as such. Beeckman even sought them out only to expose them, like the alchemist and glassblower Jacobus Bernhardi and the libertine Johannes Torrentius, whom Beeckman suspected of being a secret follower of the radical Anabaptist sect of David Jorisz (ibid., 2:201, 364–65).

34. Ibid., 3:156–57, 336; 4:263.

35. Ibid., 3:61. Beeckman explicitly refers to "natuerlicke wetenschappen ende mathematische konsten."

36. Ibid., 2:455–56: "in de philosophie niet begrepen synde theologie en politie." This sounds very much like Robert Hooke's draft of the statutes of the Royal Society, in which he stated that the purpose of the Royal Society was "to improve the knowledge of naturall things, and all useful Arts, Manufactures, Mechanick Practices, Engynes and Inventions by Experiment (not meddling with Divinity, Metaphysics, Moralls, Politicks, Grammar, Rhetorick or Logick)." Quoted in Martha Ornstein, *The Rôle of Scientific Societies in the Seventeenth Century* (Chicago: University of Chicago Press, 1913; reprint, 1938), 108–9. After the turmoil of the English Civil War, this seemed like the best way to enable members with differing political allegiances to discuss scientific topics. Perhaps Beeckman was thinking along the same lines, given the troubles he had experienced in Rotterdam.

37. *Journal*, 3:61–62. In 1628, as he was adding marginal headings to his notes, Beeckman referred to this "college" as a "Collegium physico-mathematicum."

38. The first permanent learned society in the Netherlands was the Hollandsche Maatschappij der Wetenschappen, founded in Haarlem in 1752 (and still going strong). The late appearance of learned societies in the Netherlands is usually attributed to lack of princely patronage, the political fragmentation of the country, and the modern outlook of the Dutch universities.

39. *Journal*, 4:179, 191.

40. Ibid., 3:34–35, 216.

41. Ibid., 3:216; cf. 3:214–15.

42. Ibid., 2:388–89; 3:47, 140.

43. Ibid., 3:140.

44. It is tempting to compare Van der Veen to the Italian miller Menocchio, who was put to death in 1599 because of his radical and materialistic opinions concerning God and the cosmos, and whose mental makeup has been so brilliantly described by Carlo Ginzburg, *The Cheese and the Worms: The Cosmos of a Sixteenth-Century Miller* (London: Routledge & Kegan Paul, 1980), originally published in Italian in 1976. However, in Van der Veen's case it was not

popular culture that surfaced and, in contrast to Menocchio, Van der Veen was a well-to-do and well-respected burgher. Contrary to the impression given by Cornelis de Waard, the life of Van der Veen is reasonably well documented. Between 1612 and 1657, he is mentioned as the owner of at least three mills in Gorkum, one of his brothers was the local apothecary, a second brother was a cloth merchant, and one of his sisters was married to a member of the city council of Gorkum (A. J. Busch, *Molens in Gorinchem: Wetenswaardigheden over de plaat-selijke molens in de loop der eeuwen* [Gorinchem: Mandarijn, 1978], 21, 24–26; Municipal Archives Gorkum, Notarial Archives, inv. nr. 4004). See also A. L. van Tiel, "De Gorcumse run-molenaar Balthasar van der Veen de Jonge," *Oud-Gorcum Varia* 10 (1993): 163–71. I kindly thank H. F. de Wit for these references. The French cleric and philosopher Pierre Gassendi, when visiting the Netherlands in 1629, also visited Van der Veen, no doubt on Beeckman's recommendation. In a letter to his patron Nicolas-Claude Fabri de Peiresc, he does not mention Van der Veen by name, yet he writes: "In Gorkum, there is a Maronite who has admirable opinions concerning to disposition of the Earth" (*Journal*, 4:153). In a letter to Beeckman, dated September 14, 1629, Gassendi refers to "our admirable friend Balthazar" (ibid., 4:156). The identification of Van der Veen as a Maronite is intriguing. Maronites are an Eastern-rite community within the Roman Catholic Church and in the seventeenth century lived in Lebanon (although there was a Maronite College in Rome established by Pope Gregory XIII in 1583 and run by Jesuits). Nothing is known of Maronites living in Holland, and it is highly unlikely that Van der Veen really was a Maronite. As De Wit already suggested, he probably was a Mennonite. Perhaps Gassendi was unfamiliar with this Protestant denomination and on hearing someone being called a Mennonite assumed that he was a Maronite.

45. On Colvius, see *Nieuw Nederlandsch Biografisch Woordenboek*, vol. 1, cols. 627–29; Caroline Louise Thijssen-Schoute, "Andreas Colvius, een correspondent van Descartes," in Thijssen-Schoute, *Uit de Republiek der Letteren: Elf studiën op het gebied der ideeëngeschiedenis van de Gouden Eeuw* (The Hague: Martinus Nijhoff, 1967), 67–89; G. D. J. Schotel, *Kerkelijk Dordrecht: Eene bijdrage tot de geschiedenis der vaderlandsche Hervormde Kerk, sedert het jaar 1572*, 2 vols. (Utrecht: N. van der Monde, 1841–45), 1:320–23.

46. *Journal*, 3:171, 216, 223. Much later, Colvius also showed the manuscript *Del flusso e reflusso del mare* (completed by Galileo around 1616) to Christiaan Huygens (*Nieuw Nederlandsch Biografisch Woordenboek*, vol. 1, col. 627). Klaas van Berkel, "Galileo in Holland before the Discorsi: Isaac Beeckman's Reaction to Galileo's Work," in C. S. Maffioli and L. C. Palm, eds., *Italian Scientists in the Low Countries in the XVIIth and XVIIIth Centuries* (Amsterdam: Rodopi, 1989), 101–10, esp. 104–5. Beeckman mentions Colvius for the first time on October 8, 1627, when Colvius showed him Gilbert's *De Magnete* (*Journal*, 3:17).

47. On Cats, see *Nieuw Nederlandsch Biografisch Woordenboek*, vol. 6, cols. 279–85; H. Smilde, *Jacob Cats in Dordrecht: Leven en werken gedurende de jaren 1623–1636* (Groningen: Wolters, 1938).

48. *Journal*, 1:226; 3:173. Beeckman's literary taste is discussed by Karel Bostoen, "Dingman Beens en de kamer van Vreughdendal," *Jaarboek De Oranjeboom* 34 (1981): 134–63, esp. 149–53.

49. *Journal*, 3:282. There are more indications that Beeckman was on very friendly terms with the De Witt family. He was particularly well informed about the illness of De Witt's wife, Anna van der Corput (about which more later). John de Witt, the famous son of Jacob de Witt, was born on September 24, 1625, and entered the Latin school in 1635. According to Rowen, "it was probably from Beeckman that John acquired his first taste for mathematics,"

but there is nothing to substantiate this claim. We can only speculate that in the mid-1630s Beeckman already noticed the extraordinary talents of the boy and discussed his education with his parents. John de Witt matriculated in Leiden in 1641 at the age of sixteen. Rowen, *John de Witt*, 12. Cf. *Journal*, 3:322.

50. On Van Beverwijck, see *Nieuw Nederlandsch Biografisch Woordenboek*, vol. 1, cols. 327–32; E. D. Baumann, *Johan van Beverwijck in leven en werken geschetst* (Dordrecht: Revers, 1910). For a list of curators, see Matthys Balen, *Beschryvinge der stad Dordrecht* (Dordrecht: Simon onder de Linde, 1677), 675–77. Among the curators mentioned, we also find De Witt (since 1621), Cats (since 1628), and Balthasar Lydius (from 1627 to 1629).

51. As quoted in Baumann, *Johan van Beverwijck*, 41.

52. *Journal*, 3:313; cf. ibid., 3:292, n. 51.

53. On the reception of Harvey's theory in the Dutch Republic, see T. S. Israëls and C. E. Daniëls, *De verdiensten der Hollandsche geleerden ten opzichte van Harvey's leer van den bloedsomloop* (Utrecht: Leeflang, 1883); J. Schouten, *Johannes Walaeus: Zijn betekenis voor de verbreiding van de leer van de bloedsomloop* (Assen: Van Gorcum, 1972); M. J. van Lieburg, "De ontvangst van Harvey's ontdekking in Nederland," *Nederlands Tijdschrift voor Geneeskunde* 122 (1978): 1473; Van Lieburg, "Isaac Beeckman (1588–1637) and His Diary-Notes on William Harvey's Theory on Blood Circulation (1633–1634)," *Janus* 69 (1982): 161–83. Reyer Hooykaas, "Science and Religion in the Seventeenth Century: Isaac Beeckman, 1588–1637," *Free University Quarterly* 1 (1951): 169–83, esp. 177, maintains that Beeckman learned about Harvey through Ent, but there is no proof for that statement. How Beeckman learned of the new theory is unclear. He does not seem to have read the *De motu cordis*. Perhaps he simply heard about the "sententia de circulatione sanguinis" (*Journal*, 3:292, 297, 298) and only afterward read about it in Jacobus Primrosius's attack on the theory, the *Exercitationes et animadversiones in librum De motu cordis et circulatione sanguinis* (London, 1630) (*Journal*, 3:296–98, 312–15, 321–22).

54. *Journal*, 4:127, 155, 208. The people concerned were the city physician Cornelis van Someren and the Reformed minister Gosuinus à Buytendijk.

55. Ibid., 3:63.

56. Ibid., 3:330.

57. Ibid., 1:87; 3:63, 330. See also De Waard, "Note sur le manuscrit," ibid., 1:xxviii. Maybe his "table book" was one of those erasable tables used by artists, merchants, and composers, that is, an almanac combined with blank pages that had been varnished so that texts could be removed with a sponge. See Peter Stalleybrass et al., "Hamlet's Tables and the Technologies of Writing in Renaissance England," *Shakespeare Quarterly* 55 (2004): 379–419.

58. *Journal*, 3:94.

59. In the few scattered remarks about his German adventures, Descartes never mentions Beeckman (or hardly anyone else for that matter), but, according to Cole, Beeckman figures prominently in the so-called Olympian dreams that Descartes said he had in 1619 and which led to his new philosophy. John R. Cole, *The Olympian Dreams and Youthful Rebellion of René Descartes* (Urbana: University of Illinois Press, 1992), 139–46.

60. Why Descartes really wanted to leave France is a topic of much debate. Did he want to escape the pressure put on him by the highest officials in the Catholic Church or did Descartes simply flee his family, who still wanted their good-for-nothing son and brother to become a well-respected councillor somewhere and get married? Cole, *Olympian Dreams*, 91–113; Richard A. Watson, *Cogito, ergo sum: The Life of René Descartes* (Boston: David R. Godine,

2002), 115–54. Descartes himself always stressed the peace and quiet he could find in the Dutch cities, and there is actually no reason not to believe him on this point. Descartes to Balzac, May 5, 1631, *Oeuvres de Descartes*, publié par Ch. Adam et Paul Tannery, 13 vols. (Paris: Vrin, 1964–75), 1:203–4.

61. *Journal*, 3:95.

62. Ibid., 3:109. According to Gustave Cohen, *Écrivains français en Hollande dans la première moitié du XVIIe siècle* (Paris: Librairie Ancienne Éduouard Champion, 1920), 432–34, Descartes visited Beeckman again on February 1, 1629, but it is much more likely that around that time Beeckman only received the letter from Paris wherein Descartes approved of Beeckman's solution to the mathematical query.

63. *Journal*, 3:112, 114. Whether Descartes really returned to Paris between his first and his second visit to Beeckman (that is, between October 8, 1628 and March 1629) has been the subject of some discussion in biographies of Descartes. Cohen, *Écrivains français*, 431–35, opted for an uninterrupted stay in the Netherlands (for a summary of the discussion, see G. Rodis Lewis, *L'oeuvre de Descartes*, 2 vols. [Paris: Vrin, 1971], 1:102–4). Stephen Gaukroger, *Descartes: An Intellectual Biography* (Oxford: Clarendon Press, 1995), 433, n. 1, is inclined to do the same. Yet in March 1629, Beeckman in a letter to Mersenne states very clearly that Descartes had recently gone back to Paris and then returned to Holland (Descartes being "peri-grinandi cupidus," desirous of traveling around) (*Journal*, 4:142). Also Descartes's own words in the *Discourse on the Method*, published in June 1637, suggest that he only settled down in the Netherlands in the spring of 1629. "Exactly eight years ago," he writes, he resolved to leave France and to settle in the Dutch Republic. *The Philosophical Writings of Descartes*, trans. John Cottingham et al., 3 vols. (Cambridge: Cambridge University Press, 1985), 1:126.

64. Descartes was very optimistic about the magnifying power of hyperbolic lenses. He even boasted to Ferrier that one could see letters on the moon with such lenses, a topic discussed with Beeckman on the renewal of their friendship in the fall of 1628 or the spring of 1629. *Journal*, 3:114.

65. Ibid., 4:148.

66. Howard Jones, *Pierre Gassendi, 1592–1655: An Intellectual Biography* (Nieuwkoop: De Graaf, 1981). For additional literature, see the bibliographical essay at the end of this volume.

67. F. Sassen, *De reis van Pierre Gassendi in de Nederlanden (1628–1629)* (Amsterdam: Noord-Hollandse Uitgevers Maatschappij, 1960), gives an overview of Gassendi's journey.

68. In 1628 the French coastal city of La Rochelle—the last military stronghold of the Huguenots and at that time besieged by the French king—was the place to be. Descartes reportedly visited this siege with a friend (who was an engineer). Upon witnessing how the city had been starved into surrender, Descartes seemed to lose all interest in the military.

69. *Journal*, 3:123; 4:189.

70. Ibid., 4:149–51. Parhelia were commonly, like rainbows, thought to be useful for predicting the weather, but they were also, again like the rainbow, imbued with symbolical meaning. Parhelia are formed around the sun when the sun's rays are reflected through a thin cloud of ice crystals, resulting in the appearance of at least one circle with a red exterior and a blue interior. Kepler had already written about parhelia in 1604–5, explaining their true nature, but the phenomenon was still widely discussed around 1630. Upon receiving a copy of Scheiner's report, Descartes took the study of the parhelia as the point of departure for the exposition of his natural philosophy in the unpublished *Le Monde*. Gaukroger, *Descartes*, 217.

71. "Dedi ei Corollaria mea" (*Journal*, 3:123). The copy of the *Theses* found in the British Museum contains a handwritten additional corollary on the corpuscular nature of sound: "The matter of the sound that reaches the ear of the listener is the very same air that was in the mouth of the speaker" (Soni materia quae aures ingreditur auditurus motura est ille idem numero aer, qui erat in ore loquentis). (For a discussion of the translation of *numero* by "the very same," see H. Floris Cohen, *Quantifying Music: The Science of Music at the First Stage of the Scientific Revolution, 1580–1650* [Dordrecht: Reidel, 1984], 275, n. 22.) This corollary does not figure on the list of provisional theses from August 1618 (*Journal*, 1:200–201), but the topic was extensively discussed during Gassendi's visit to Beeckman in 1629. Actually, just before Beeckman notes that he has given a copy of his corollaries (note that he does not use the term "thesis" here) to Gassendi, he touches on the nature of sound in almost the same words as in the handwritten corollary: "Dixi etiam aerem, qui auditum movet, esse eundem numero qui erat in ore loquentis." Therefore, it is possible Beeckman gave the mutilated copy of the thesis (actually only the corollaries) to Gassendi; Beeckman might have added a new corollary for the occasion. One can only speculate whether Beeckman tore out the central pages of his thesis because he no longer wanted a philosopher like Gassendi to read his outdated theses on malaria.

72. The city surrendered to Frederick Henry on September 14, 1629.

73. *Journal*, 4:153.

74. Ibid., 4:189.

75. Ibid., 3:98.

76. On Rivet, see *Nieuw Nederlandsch Biografisch Woordenboek*, vol. 7, cols. 1051–52; H. J. Honders, *Andreas Rivetus als invloedrijk gereformeerd theoloog in Holland's bloeitijd* (The Hague: M. Nijhoff, 1930); A. G. van Opstal, *André Rivet: Een invloedrijk Hugenoot aan het hof van Frederik Hendrik* (Harderwijk: Flevo, 1937). Rivet became governor to William II, the son of stadholder Frederick Henry in 1636, after having lectured in Leiden for sixteen years. On Rivet and the Republic of Letters, see J. A. H. Bots, "André Rivet en zijn positie in de Republiek der Letteren," *Tijdschrift voor Geschiedenis* 84 (1971): 24–35; Paul Dibon, *Inventaire de la correspondance d'André Rivet (1595–1650)* (The Hague: Martinus Nijhoff, 1971). Just as Rivet established the contact between Beeckman and Mersenne, so Beeckman was instrumental in bringing together Rivet and Descartes, who then was introduced to Henricus Reneri, soon to become one of Descartes's most loyal friends in the Netherlands. As professor of philosophy at the illustrious school in Deventer and the University of Utrecht, Reneri is said to be the first to lecture on Cartesian ideas in Dutch universities, but there is little to substantiate this claim. Reneri also was Beeckman's friend, at least since 1626 (*Journal*, 2:371). See the article by Theo Verbeek in Wiep van Bunge et al., eds., *Dictionary of Seventeenth- and Eighteenth-Century Dutch Philosophers*, 2 vols. (Bristol: Thoemmes Press, 2003), 2:824–26.

77. *Correspondance du P. Marin Mersenne religieux minime*, 17 vols. (Paris: Beauchesne, 1932–88), 2:103.

78. *Journal*, 4:133.

79. Ibid., 4:141–44.

80. On Mersenne, see Robert Lenoble, *Mersenne ou la naissance du mécanisme* (Paris: Vrin, 1943, reprint, 1971 and 2003); Peter Dear, *Mersenne and the Learning of the Schools* (Ithaca: Cornell University Press, 1988).

81. Lenoble, *Mersenne*, 2: "secrétaire général de l'Europe savante."

82. Ferd. Sassen, *De reis van Mersenne in de Nederlanden (1630)* (Brussels: Koninklijke Vlaamse Academie voor Wetenschappen, Letteren en Schone Kunsten van België, 1964), gives a good survey of the trip through the Netherlands. Mersenne did not travel alone; he (probably) was accompanied by a servant. See *Correspondance de Mersenne*, 2:525.

83. *Journal*, 3:160−64.

84. At Beeckman's quarters, Mersenne also met Johan van Beverwijck and Van Beverwijck's brother-in-law, Jan Oem van Wyngaerden, president of the Court of Holland (the highest court of justice in the province) (ibid., 4:213−14).

85. Klaas van Berkel, "Descartes' Debt to Beeckman; Inspiration, Cooperation, Conflict," in Stephen Gaukroger et al., eds., *Descartes' Natural Philosophy* (London: Routledge, 2000), 46−59.

86. *Journal*, 4:133.

87. Ibid., 4:133, n. 7.

88. Beeckman to Mersenne, March 1629, ibid., 4:141−42.

89. Watson, *Cogito, ergo sum*, 87.

90. *Journal*, 4:163.

91. Ibid., 4:194.

92. "De dire que la mesme partie d'air, in individuo, qui sort de la bouche de celuy qui parle, va fraper toutes les oreilles, cela est ridicule" (ibid., 4:177). In a note De Waard points out that Descartes gives a distorted picture of Beeckman's theory. Cohen disagrees. See Cohen, *Quantifying Music*, 275, n. 22 (for a further discussion, see chapter 4).

93. Descartes to Beeckman, December 18, 1630, *Journal*, 4:171.

94. Descartes to Beeckman, February 25, 1630, ibid., 4:179.

95. Descartes to Mersenne, August 1630, ibid., 4:192.

96. Ibid., 4:191−92, 195.

97. Ibid., 4:192.

98. Ibid., 4:194−95. It is true that we are not able to answer the question whether Descartes's 1619 letters to Beeckman were really extravagantly praising his friend, because none of Descartes's letters from that time to others exist, but we do know that later on Descartes never wrote letters excessively praising his correspondents (or anyone else).

99. Ibid., 4:195. The letter itself has not been preserved. It is only through Descartes's citing of this letter in his letter to Mersenne at the end of October 1630 that we know that Beeckman had written it.

100. This was Descartes's usual way of putting down Beeckman. In the same letter, Descartes denied that Beeckman even understood what a hyperbola is, "except perhaps as a petty teacher of grammar" (nisi forte tanquam Grammaticulus) (ibid., 4:199). See also chapter 5, for further instances of Descartes's scornful remarks on Beeckman as a simple schoolteacher.

101. Ibid., 4:196.

102. Ibid.

103. Ibid., 4:198. On Descartes's typology of discoveries, see W. A. Gabbey "Descartes' Dynamical Thought. A critical Study of Some Problems, Principles and Concepts" (Ph.D. dissertation, Queens University Belfast, 1964), 545−51; Gabbey "Les trois genres de découverte selon Descartes," *Actes du XIIe Congrès International d'Histoire des Sciences. Paris 1968* (Paris, 1968), 2:45−49. Elsewhere in the letter, Descartes tried to refute two claims made by Beeckman in more detail. The first concerned Beeckman's proof that pitch is inversely pro-

portional to the length of a cord (which, according to Descartes, could be found in Aristotle). The second concerned the proof that a hyperbolic lens gathers all incoming rays in one point (according to Descartes a simple extension of his own proof for elliptic lenses). Descartes was wrong in both cases (*Journal*, 4:199). On the first point, see also chapter 5.

104. Cohen, *Quantifying Music*, 196.

105. Gaukroger, *Descartes*, 224.

106. *Journal*, 3:74. Cf. John Schuster's Princeton dissertation *Descartes and the Scientific Revolution, 1618–1634: An Interpretation* (Ann Arbor: University Microfilms International, 1977), 566–79.

107. Descartes to Mersenne, October 8, 1629, *The Philosophical Writings of Descartes*, 3:6. In Descartes's little treatise, which eventually would evolve into his *Le Monde*, he explained a number of natural phenomena on the basis of the micromechanical philosophy he had first come across in Beeckman's notebooks. In 1633 Descartes decided not to publish the manuscript after he had heard the news about Galileo's condemnation.

108. Van Berkel, "Descartes' Debt to Beeckman." This is also the interpretation of Schuster, *Descartes*, 590–93. There is an instructive parallel with the case of Ferrier, the lens grinder with whom Descartes worked on the hyperbolic lenses in the late 1620s. When Ferrier proved to be too self-conscious and not completely prepared to accept Descartes as his superior in practical optics around 1630, Descartes reacted vehemently by denigrating Ferrier as a nearly illiterate artisan who would never be able to accomplish anything without Descartes's theoretical guidance. Watson, *Cogito, ergo sum*, 159.

109. Descartes to Etienne de Ville-Bressieu, August or September 1631, *Journal*, 4:205; Beeckman to Mersenne, October 7, 1631, ibid., 4:206–7.

110. Descartes to Beeckman, August 22, 1634, ibid., 4:225–28.

111. Ibid., 3:210.

112. Ibid., 1:87.

113. Ibid., 4:263.

114. Ibid., 1:xxix. Another argument against De Waard's hypothesis is the fact that the extant notebooks contain Beeckman's notes about his loss of weight and his health in general up till the end of his life.

115. "Etiam nunc incomptus sum vestibus, ac ordinationis bibliothecae et musei negligentissimus" (ibid., 3:220).

116. Ibid.

117. Unfortunately we have no portrait of Beeckman to help us fathom his personality. We have to rely on what he told us himself in his *Journal*: 2:307; 3:219, 220, 220–21.

118. Ibid., 3:63.

119. Beeckman's children were:

> 1. Jacob, born March 7, 1621, died after a few weeks (ibid., 2:163; 4:73). In the notebooks, f. 150r., Beeckman explains why he called his son Jacob, namely "because I am called Isaac and my father Abraham, as a reference to the three patriarchs, and [finally] as a way to better remember our genealogy."
>
> 2. Nameless boy, born November 11, 1622, died the next day (ibid., 4:74).
>
> 3. Catharine or Catelyntje, born March 27, 1624, buried June 8, 1708 (ibid., 1:xxiii, n. 12; 4:81).

4. Jacob, born September 4, 1627, died October 22, 1628 (ibid., 4:127).

5. Jacob, born August 14, 1629, died July 4, 1631 (ibid., 4:155, 205).

6. Abraham, born February 13, 1632, died August 8, 1632 (ibid., 3:238, n. 1; 4:208).

7. Susanna, born October 28, 1633, buried October 13, 1638 (ibid., 4:218, 284).

120. Ibid., 2:388.

121. Ibid., 4:205–8.

122. Ibid., 3:210.

123. Ibid., 4:232.

124. Ibid., 4:369.

125. Baumann, *Johan van Beverwijck*, 44–50. The first notes in Beeckman's *Journal* related to the plague date from May–June 1634 (*Journal*, 3:351).

126. Van Bos, *Dordrechtsche Arcadia*, 575.

127. *Journal*, 3:369; 4:229.

128. Ibid., 4:230.

129. On the illustrious schools in general, see Frijhoff, *La société néerlandaise*, 15–17, 279–81; on the schools in Amsterdam, Breda, Deventer, Harderwijk, and Rotterdam, see the entries in Van Bunge et al., *Dictionary of Seventeenth- and Eighteenth-Century Dutch Philosophers*. Another reason for establishing an illustrious school in Dordrecht may have been the wish to outdo Rotterdam, where in 1634 the magistrates also planned to institute such a school. See *Florislegium: Bloemlezing uit de Erasmiaanse, Rotterdamse en andere opstellen van drs. N. van der Blom* (Leiden: Brill / Backhuys, 1982), 64. Whether Beeckman was consulted during the preparations of the decision to establish an illustrious school cannot be determined from the *Journal*, although he was in Rotterdam in January 1634 (*Journal*, 3:333). As such, an illustrious school did resemble his own ideas of the perfect academy, as can be deduced from a remark made in August 1634 (ibid., 3:354). He was also acquainted with the workings of the illustrious school at Utrecht. Shortly after the opening of this school (June 28, 1634), he had visited Utrecht and had talked to two of the recently appointed professors, Antonius Aemilius and Henricus Reneri, both of them old friends (ibid., 3:353, 354, 384).

130. *Journal*, 4:212.

131. Ibid., 4:231, n. 6. Walen was a brother-in-law of Johan van Beverwijck and in September 1636 succeeded him as physician to the city. See Baumann, *Johan van Beverwijck*, 50. The illustrious school in Dordrecht was officially opened in the fall of 1635 (ibid., 46–47).

132. It is difficult to ascertain how many people died from the plague in Dordrecht, but in Leiden (where health conditions were perhaps worse than in Dordrecht) the plague killed 14,582 people from June 23 to December 31, 1635, more than 500 per week.

133. On the consequences of the plague in Dordrecht, see Baumann, *Johan van Beverwijck*, 47–50.

134. *Journal*, 2:294.

135. Ibid., 2:295. Rolf Willach, *The Long Route to the Invention of the Telescope* (Philadelphia: American Philosophical Society, 2008), amply demonstrated the poor quality of most of the extant seventeenth-century lenses.

136. *Journal*, 3:44–45, 69. Books consulted by Beeckman include Johannes Kepler, *Dioptrice* (Augsburg, 1611); Franciscus Aguilonius, *Opticorum libri sex* (Antwerpen, 1613); and Hieronymus Sirtorius, *Telescopium sive Ars proficiendi novum illud Galilaei visorium instrumentum ad Sydera* (Frankfurt, 1618).

137. *Journal*, 3:215, 396.

138. Lipperhey applied for a patent and was sent to The Hague to show his "instrument to see far" to the States General. He also demonstrated the invention to Prince Frederic Henry and the Spanish general Spinola, who was in The Hague for negotiations regarding a truce. Eventually, Lipperhey did not manage to secure a patent, because others also claimed to have invented the instrument. The demonstration to Frederic Henry and Spinola, and the stir this generated all over Europe inspired Galileo to make a telescope himself. See Rienk Vermij, "The Telescope at the Court of the Stadtholder Maurits," in Albert van Helden et al., eds., *The Origins of the Telescope* (Amsterdam: KNAW Press, 2010), 73–92.

139. *Journal*, 3:249, 376–77. Sachariassen was the son of the spectacle maker (and counterfeiter) Sacharias Jansen, who until the beginning of this century was still regarded by some as the true inventor of the telescope. This claim was partially based on Sachariassen's boasting to Beeckman that his father had been the first in the Dutch Republic to copy the telescope from some unknown Italian (ibid., 3:376). Recent research by Rolf Willach and Huib Zuidervaart has once and for all disproved this claim. Lipperhey is the true inventor of the telescope, in the sense that he was the first one to use a diaphragm. (Beeckman discusses the diaphragm in 1618 in Caen, stating that he remembers meeting a lens grinder in Middelburg who thought he could do *without* a diaphragm [ibid., 1:209]. He must therefore have been familiar with the Middelburg lens grinders.) See Willach, *The Long Route*; Huib Zuidervaart, "The 'True Inventor' of the Telescope: A Survey of 400 Years of Debate," in Van Helden, *The Origins of the Telescope*, 9–44.

140. *Journal*, 3:308, 383, 389.

141. Ibid., 3:391.

142. Ibid., 3:386.

143. Ibid., 3:430.

144. Ibid., 3:158.

145. Ibid., 3:203–4, 442. It is not known how Beeckman got the copy of Drebbel's letter. Beeckman's father, who had sometimes provided him with news about Drebbel's inventions, had died in 1625.

146. Ibid., 3:358; 4:232–33. Cf. J. van der Hoeven, "Een brief van Justinus van Assche aan Isaac Beeckman," *Bijdragen tot de Geschiedenis der Geneeskunde* 13 (1933): 17–23 (a letter of Van Assche to Beeckman, dated July 30, 1636, not included in De Waard's edition of the *Journal*, which deals with the plague and with family business).

147. *Journal*, 2:199; 3:86, 263, 265. On Blaeu, see P. J. H. Baudet, *Leven en werken van Willem Jansz. Blaeu* (Utrecht: Van der Post jr., 1871); *Nieuw Nederlandsch Biografisch Woordenboek*, vol. 10, cols. 74–78; J. Keuning, M. Donkersloot-de Vrij, *Willem Jansz. Blaeu: A Biography and History of His Work as a Cartographer and Publisher* (Amsterdam: Theatrum Orbis Terrarum, 1973); Djoeke van Netten, *Koopman in kennis: De uitgever Willem Jansz. Blaeu (1571–1638) in de geleerde wereld van zijn tijd* (Groningen: printed by the author, 2012).

148. E. W. Moes, "Martinus Hortensius, de eerste hoogleeraar in de mathematische wetenschappen in Amsterdam," *Oud-Holland* 3 (1885): 209–16; *Nieuw Nederlandsch Biografisch Woordenboek*, vol. 1, cols. 1160–64; Klaas van Berkel, "De illusies van Martinus Hortensius: Natuurwetenschap en patronage in de Republiek," in Van Berkel, *Citaten uit het boek der natuur: Opstellen over Nederlandse wetenschapsgeschiedenis* (Amsterdam: Bert Bakker, 1998), 63–84; Annette Imhausen and Volkert R. Remmert, "The Oration on the Dignity and the Usefulness of the Mathematical Sciences of Martinus Hortensius (Amsterdam 1634): Text,

Translation and Commentary," *History of Universities* 21 (2006): 71–150 (Hortensius had given his inaugural lecture on May 8, 1634). Perhaps we should include George Ent among Beeckman's students who remained interested in and were active in natural philosophy. Ent—who in 1627 had gone to see Beeckman to deepen his knowledge of natural philosophy—published his *Apologia pro circulatione sanguini* in 1641. In the *Apologia*, Ent defended Harvey's theory with arguments taken from Beeckman and not, as is commonly assumed, from Descartes. See Van Lieburg, "Isaac Beeckman and His Diary-Notes on Harvey."

149. "Vidit et cum judicio percurrit librum hunc meditationum mearum" (*Journal*, 3:354). On receiving the *Dialogo*, Beeckman immediately took extensive notes. Later that month he lent the book to Descartes for two days. Descartes to Mersenne, August 14, 1634, ibid., 4:224–25.

150. For a survey of the problem of longitude, see E. Crone, "Introduction [to Stevin's *Haven-finding Art*]," in *The Principal Works of Simon Stevin*, ed. Ernst Crone et al., 5 vols. (Amsterdam: C. V. Swets and Zeitlinger, 1955–65), 3:363–417; C. A. Davids, *Zeewezen en wetenschap: De wetenschap en de ontwikkeling van de navigatietechniek in Nederland tussen 1585 en 1815* (Dieren: De Bataafsche Leeuw, 1986); Dava Sobel, *Longitude: The True Story of a Lone Genius Who Solved the Greatest Scientific Problem of His Times* (London: Fourth Estate, 1995).

151. Stillman Drake, *Galileo at Work: His Scientific Biography* (Chicago: University of Chicago Press, 1978), 193–94, 257–61; G. Vanpaemel, "Science Disdained: Galileo and the Problem of Longitude," in Maffioli and Palm, *Italian Scientists in the Low Countries*, 111–29; Silvio A. Bedini, *The Pulse of Time: Galileo Galilei, the Determination of Longitude, and the Pendulum Clock* (Florence: Leo S. Olschki, 1991); J. L. Heilbron, *Galileo* (Oxford: Oxford University Press, 2010), 346–48.

152. *Journal*, 4:235. Reael (1583–1637) had studied law at Leiden University and had defended his dissertation in 1608. From 1611 until 1620 he had been in the East Indies, from 1616 to 1619 as governor-general. After his return, he had some difficulty in finding an official position, mainly because of his Remonstrant persuasion, but after the death of Prince Maurice in 1625 he again fulfilled several high commissions. In 1630 he became a member of the Amsterdam magistrate. Reael's main interest was navigation. *Nieuw Nederlandsch Biografisch Woordenboek*, vol. 4, cols. 1121–25. Descartes's biographer Adrien Baillet mentions that he "had the reputation of being the number one of his age in the philosophy of magnetism" (passoit pour le premier homme du siècle dans la philosophie magnétique). His *Observatien of ondervindingen aen de magneetsteen: Quibus adjunctae sunt . . . Prof. Casparis Barlaei Causae et Rationes Observationum earumdem magneticarum* (Amsterdam, 1651) was published posthumously. See also *Correspondance de Mersenne*, 2:580.

153. *Journal*, 4:236.

154. Ibid., 4:236–37. Cf. Stephane Garcia, *Élie Diodati et Galilée: Naissance d'un réseau scientifique dans l'Europe du XVIIe siècle* (Florence: Leo S. Olschki, 2004).

155. *Journal*, 4:241–44; Drake, *Galileo at Work*, 374.

156. *Journal*, 4:245–51.

157. Ibid., 4:253.

158. Ibid., 3:229–30. Earlier, in 1614, Beeckman had thought of the moon as the "heavenly clock" (hemelse klok), see ibid., 1:33–34; 4:62. Beeckman's official nomination has not been found. Presumably he was admitted to the commission on the same informal basis as Golius. Cf. ibid., 4:282, 288.

159. Ibid., 4:261–62.

160. Ibid., 4:267–68.

161. Ibid., 4:267.

162. Ibid., 4:270.

163. Ibid., 4:258–59.

164. *Le opere di Galileo Galilei*, ed. Antonio Favaro, 20 vols. (Florence, 1892–1909; reprint, Florence: Barbèra, 1968), 17:96–105, 174–75.

165. *Journal*, 4:282, n. 1.

166. *Correspondance de Mersenne*, 9: 172; Baudet, *Blaeu*, 22.

167. *Journal*, 4:155; 3:443–50.

168. Ibid., 4:431, n. 1.

169. Ibid., 4:263.

170. Ibid., 3:431, n. 1.

171. Balen, *Beschryvinge der stad Dordrecht*, 677, gives May 20 as the date of his death, whereas Abraham Beeckman in the *Journal* gives the date as May 19 (*Journal*, 1:xxii). Abraham's entry is more trustworthy, because he was very close to his brother during his last years.

172. Descartes to Colvius, June 14, 1637. *Journal*, 4:281–82 (translation Desmond M. Clarke). At that moment, Descartes was in Leiden to oversee the publication of his *Discourse on the Method*.

173. Cole, *Olympian Dreams*, 90.

174. *Journal*, 4:283.

175. Ibid., 4:282.

176. Ibid., 4:283.

177. Ibid., 4:284. Her mother, Beeckman's widow, died soon after, in 1638.

178. Ibid., 4:286.

179. Eugenio Canone, "Il *Catalogus librorum* di Isaac Beeckman," *Nouvelles de la République des Lettres* 1991, no. 1 (1992): 131–59, with a reproduction of the catalog. Its full title runs as follows: *Catalogus Variorum et insignium Librorum Clarissimi Doctissimique viri D. Isaaci Beeckmanni, Praestantissimi Medici, Philisophi [sic] atque Mathematici autissimi [sic], Schelae [sic] Durdracaenae Rectoris vigilantissimi. Quorum Auctio habebitur in aedibus defuncti ad diem 1[4] Julij MDCXXXVII.* [vignette] Durdrechti. Typis Isaaci Andreae MDCXXXVII. The only extant copy is in the Biblioteca Angelica in Rome.

180. Missing is also the copy of Girard Desargues's *Exemple de l'une des manieres universelles de S.G.D.L. touchant de la pratique de la perspective* (Paris, 1636), which the author sent to Beeckman with a handwritten dedication, "Pro viro Clarissimo Isaac Beeckman Dordracensis Collegii Rectore" (*Journal*, 4:240–41). The engineer, mathematician and architect Girard Desargues (Lyon, 1591–Paris, 1661) was a close friend of Mersenne.

181. Also in this case, at the end of the catalog, the bookseller mentions "some other parcels" of unspecified books, probably small books used for teaching.

182. The catalog is divided in three headings, for theology, for medicine, and for what is called "Philosophi, historici etc," each category subdivided according to size (folio, quarto, octavo, and the smaller sizes). Identifying a book as "general philosophy" or "natural philosophy" is therefore sometimes open to question, as for instance with regard to Aristotle's *Opera graece*, which of course includes his works on natural philosophy. What I

clustered as "mathematics" includes books on navigation (*De schatkamer des grooten Zee-vaerts*, in folio—as yet unidentified), on architecture (Vitruvius's *De architectura* in a 1586 edition), on elementary counting (Bartjes's *Cijferboeck*), and on music (Henricus Glarea-nus, *Dodekachordon sive: De musica opus* [1547], but also Proclus, *De sphaera*. "Natural phi-losophy" includes Bacon's *Historia naturalis*, Fludd's *Clavis philosophiae et Alchymiae*, Go-clenius's *Scholae seu disputationes physicae* (1602), and Franciscus Titelmans's *Philosophia naturalis* (1557). "General philosophy" includes Zabarella's *Opera* (Cologne, 1603), Aristo-tle's *Metaphysica* with a commentary by Fonseca (1599), Alsted's *Theatrum scholasticum* (1620), and Rudolph Snellius's commentary on Ramus's *Ethica*. The catalog lists no less than seven copies of Antonius Walaeus, *Compendium ethicae Aristotelicae ad normam veritatis Christianae revocatum* (editio auctior, Leiden, 1626), probably because Beeckman used this book in his classes.

183. As to Galen, Beeckman owned only a copy of the *Epitome operum Galeni* by Andres de Laguna (1604). The catalog contains several books from Paracelsus (and from Paracelsian au-thors and his critics), but Paracelsus is not even once mentioned in the *Journal*.

184. *Journal*, 4:283–84.

185. The only copy of the booklet known to De Waard can be found in the University Li-brary of Leiden University, sign. 534 F 9: 1 (it once belonged to Isaac Vossius). Kubbinga found at least two more copies, in Lübeck and Bonn (H. H. Kubbinga, "Nova Beeckmaniana," *De Zeventiende Eeuw* 9 [1993]: 82–83). The booklet is dedicated to Jacob de Witt, Cornelis van Beveren, Johan van Beverwijck, Mattheus Berck, Cornelis van Someren, Nicolaas Ruysch, and Gosuinus à Buytendyck, all curators of the Latin school in Dordrecht. A comparison be-tween the content of the *Journal* and the *Centuria* has been made easy because De Waard at-tached an asterisk to the notes in the *Journal* that found their way into the *Centuria*. There are no notes included in the *Centuria* after 1629, one of the last being a note concerning the visit of Gassendi in July 1629 (*Journal*, 3:123). Abraham replaced Beeckman's rather crude drawings with neat pictures. Of the hundred entries, thirty-two are in Dutch. The distribu-tion of the notes is very uneven. More than seventy date from before March 1615; the others are from the period 1627–29.

186. In some cases, Abraham Beeckman corrected evident mistakes or altered the words in order to make the meaning more general. For instance, in a note written in reaction to Gil-bert's *De magnete*, Isaac Beeckman had written: "Existimo enim stellas in Terram magnetem immittere spiritus corporeos" (I think that the stars radiate corporeal spirits to the mag-netic Earth), which Abraham Beeckman changed by adding *seu* in the phrase "in Terram seu magnetem" (in the Earth or magnet). *Journal*, 3:17, note a.

187. *Journal*, 1:xxiii (ms. notebooks, f. 296 recto).

CHAPTER 4: PRINCIPLES OF MECHANICAL PHILOSOPHY I: MATTER

1. *Journal*, 1:132–33; 3:123; 4:41.

2. Ibid., 3:112.

3. Ibid., 1:xxiii: "Is altoos besig geweest met speculeren." In early modern Dutch, the word *speculation* might be translated as "spiegheling," which in Simon Stevin's work was equivalent to "theorizing."

4. See also John Schuster, *Descartes and the Scientific Revolution, 1618–1634: An Interpreta-tion* (Ann Arbor: University Microfilms International, 1977), 57–58, for some valuable re-marks on "Beeckman's style of natural philosophical speculation."

5. *Journal*, 1:244 (a note made in the context of Beeckman's conversations with Descartes in 1618); ibid., 4:186 (in a letter to Mersenne, April 30, 1630). In both cases, the terms refer to basic explanatory principles, not to methodological principles.

6. The first explicit reference to this concept is to be found in the oration Beeckman delivered on accepting the post of principal of the Dordrecht Latin school in 1627: "Vera philosophia, id est mathematico-physica" (*Journal*, 4:126). In 1618 Descartes praised Beeckman because he so "accurately unified physics and mathematics" (accurate cum Mathematica Physicam jungat). In 1628 Beeckman added the marginal note: "Physico-mathematici paucissimi" (ibid., 1:244).

7. Ibid., 4:40.

8. Ibid., 2:243–47. Beeckman also said that Basson would not have criticized Democritus's theory of vision "if he had studied the nature of eye tubes and burning mirrors in a proper mathematical way, like Kepler did" (si tuborum ocularium et speculorum urentium naturam *satis mathematicè*, ut Keplerus, cognovisset). On Basson, see Christoph Lüthy, "Thoughts and Circumstances of Sébastien Basson: Analysis, Micro-History, Questions," *Early Science and Medicine* 2 (1997): 1–73. According to Lüthy, Basson's book was "the earliest atomist textbook of natural philosophy."

9. *Journal*, 2:56. Cf. ibid., 3:51–52, contrasting Bacon and Stevin.

10. Ibid., 1:244.

11. Ibid., 1:54–55.

12. Ibid., 4:122–26.

13. Ibid., 3:122.

14. Ibid., 3:124.

15. Ibid., 1:170–71, 285; 2:243.

16. E. J. Dijksterhuis, *The Mechanization of the World Picture* (Oxford: Oxford University Press, 1961), 405–6.

17. In his analysis of their deduction of the law of falling bodies, Koyré stressed the contrast between Beeckman the physicist and Descartes the mathematician. Alexandre Koyré, *Études galiléennes* (Paris: Hermann, 1939; reprint, 1966), 107–36. On Beeckman, Koyré wrote: "N'oublions pas en effet, que Beeckman, bon physicien sans doute, est un mathématicien fort médiocre" (ibid., 122). Koyré's assessment of Descartes has been questioned by Schuster, *Descartes*, 84–93, but there is no reason to do the same with regard to Beeckman. He was indeed a mediocre mathematician.

18. *Journal*, 4:186.

19. This is in no way contradicted by Beeckman's reaction to Descartes's attempt, successful in the end, to stop him from publishing his physico-mathematical speculations. Beeckman did not give up his ideas; he only refrained from publishing them.

20. *Journal*, 4:184: "There is no need to find out whether the wooden globe falls slower than the iron globe, because reason again and again tells this quite clearly" (Non opus est experiri an, ligneus globus ferreo tardius cadat, cum ratio id satis manifesto dictitet).

21. Ibid., 4:217.

22. Ibid., 1:100–101.

23. Ibid., 2:101–2: "Qui tamen generalem quendam processum a primis principiis usque ad ultima composita potest sibi comparare, compendiosius totum cursum suum absolvet, omnia revocans ad suum proprium locum, utens commoda quadam rerum omnium methodo, non aliter quam ii qui imaginibus et locis fictis ipsas res suae memoriae infigunt."

24. See Frances A. Yates, *The Art of Memory* (Chicago: University of Chicago Press, 1966). The word *compendiosius* is a reference to Ramist didactics.

25. *Journal*, 4:41: "Nullum enim statutum, nullum praeceptum, nulla regula in philosophia admittenda quae non sit apodictica et certissima ratione comprobata et intellectui tam aperte et nude objecta atque visibilia oculis objiciuntur." Of course, this is reminiscent of Descartes's famous dictum that in philosophy everything had to be presented "clare et distincte." See Descartes, *Meditationes in prima philosophia* (1641), in *Oeuvres de Descartes*, publiées par Charles Adam et Paul Tannery, 13 vols. (Paris: Vrin, 1964–75), 7:69.

26. *Journal*, 4:162. The word *sensilis* is not standard Aristotelian terminology but is derived directly from Lucretius, *De rerum natura*, 2.888. See also Beeckman's criticism of Bacon's concept of folding and unfolding matter (postulated to avoid the need of a vacuum): "[These folds] cannot, in my opinion, be observed by the human mind and should therefore be expelled from the true philosophy" (Nequeunt, meo juditio, ab ingenio humano percipi; ideoque exterminandae a vera philosophia (*Journal*, 2:254).

27. *Journal*, 3:34: "Deo tribuens universae naturae compositionem, ita ut nostris ingeniis subjecerit integram contemplationem de rebus inferioribus per naturam factis."

28. Ibid., 3:18.

29. "Conformatrix natura in universo, aut potius in calore, mihi videtur nimis ridicula et philosopho indigna. Hoc enim non est causam proferre, . . . sed eam occultare" (ibid., 3:34).

30. Ibid., 3:67: "Qui enim immateriatum movere possit materiatum?"

31. Ibid., 1:121: "Ist niet vremt, dat men handen, voeten ende leden roeren kan, als men wil, deur gedachten?"

32. Ibid., 1:151: "cum omnis vis fiat contactu."

33. Ibid., 2:229, 339.

34. Ibid., 1:132: "Bodies have two attributes: first form, second motion, and rest. In the beginning God both created and moved the atomic bodies. Once moved, they never came to rest, unless impeded by each other. Therefore, combined with the interstitial vacuum, these constitute matter and form of the composite bodies in heaven and Earth" (Corpori duo attribuantur: figura primo, secundo motus et quies. Deus corpora atoma primo movit non minus quam creavit; motis semel nunquam quiescebant, nisi ab invicem impeditis. Ergo congredientes et cum vacuo misto, convenienter materia et forma extiterunt omnium compositorum coeli et Terrae etc.). Note that Beeckman here speaks only of the vacuum intermixtum, the small pockets of empty space between the particles of matter.

35. Ibid., 1:216: "Omnes igitur vires emergunt ex motu, figura et quantitate, ideoque in unaquaque re tria haec sunt consideranda."

36. Ibid., 3:310: "Quis enim comprehenderit quid sit illud quod resistentiam causatur, et unde primus motus?"

37. Ibid., 1:132: "No doubt coldness and heat are caused by the speed and slowness of small particles" (Nec dubium est etiam a celeritate et tarditate motus corpusculorum rationem frigoris et caloris sumi).

38. Dijksterhuis, *Mechanization*, 547.

39. As quoted by Margaret Osler, "How Mechanical Was the Mechanical Philosophy? Non-Epicurean Aspects of Gassendi's Philosophy of Nature," in Christoph Lüthy et al., eds., *Medieval and Early Modern Matter Theories* (Leiden: Brill, 2001), 423–39, quotation on 423–24. According to Osler, Boyle referred to the mechanical philosophy for the first time in his

Essays of 1661, a work that was written around 1655. A few years later, Robert Hooke in his *Micrographia* (1664) used the phrase "the real, the mechanical, the experimental philosophy" (as quoted in Jim Bennett, "The Mechanic's Philosophy and the Mechanical Philosophy," *History of Science* 24 [1986]): 1–28, quotation on p. 1). Christoph Lüthy pointed out to me that Henry More also used the phrase "mechanical philosophy" in his treatise *The Immortality of the Soul* (1659), referring specifically to Descartes's philosophy.

40. Alan Gabbey, "What Was 'Mechanical' about the Mechanical Philosophy," in Carla Rita Palmerino and J. M. M. Thijssen, eds., *The Reception of Galilean Science of Motion in the Seventeenth Century* (Dordrecht: Reidel, 2004), 11–23, esp. 12–13.

41. *Journal*, 1:218.

42. "Pinguiuscula" has also the connotation "greasy," which may have been implied in Froidmont's use of the word: machines need oil, and oil makes dirty hands.

43. As quoted in Gabbey, "What Was 'Mechanical,'" 18–20, quoting Descartes to Plemp for Froidmont, September 13, 1637 (cf. *The Philosophical Writings of Descartes*, trans. John Cottingham et al., 3 vols. [Cambridge: Cambridge University Press, 1985], 3:64).

44. *The Philosophical Writings of Descartes*, 3:64.

45. *Journal*, 4:25–31. Van Laren was born in Arnemuiden in 1590 and may have met Isaac Beeckman at the Latin school of Biesius. In 1610 he went to study divinity in Leiden and in 1612, he moved on to Franeker, with a couple of the students from Zeeland. A certain Jodocus van Laren from Arnemuiden, probably a brother, was already studying divinity at Franeker. Given the fact that Franeker at that time was a hotbed of heterodox opinions, including the atomism that inspired David Gorlaeus, perhaps after all there is a link between Beeckman and Gorlaeus through Jeremias van Laren (who himself was a staunch defender of Aristotelian philosophy but must have heard other students discuss atomism).

46. Ch. B. Schmitt, "Experimental Evidence for and against a Void: The Sixteenth-Century Arguments," *Isis* 58 (1967): 352–66; Schmitt, "Changing Conceptions of Vacuum (1500–1650)," in *Actes du XIème Congrès International d'Histoire des Sciences. Varsovie-Cracovie 1965* (Wroclaw-Varsovie-Cracovie, 1968), 3:340–43; Edward Grant, *Much Ado about Nothing: Theories of Space and Vacuum from the Middle Ages to the Scientific Revolution* (Cambridge: Cambridge University Press, 1981). The argument Beeckman used to prove that a vacuum does indeed exist is exactly the same argument that later troubled Descartes, who like Van Laren, but for different reasons, denied the possibility of a vacuum.

47. *Journal*, 4:27: "Pergis argumentare: In continuo aut undique contiguo, nihil movetur. In aere autem aves, in aqua pisces moventur; ergo non sunt undique contigua." This kind of arguing (using syllogisms) was something Beeckman, in later years, never practiced again. In his discussion with Van Laren, Beeckman models his logic on Keckermann's highly popular *Systema Logicae*, which is not listed in the auction catalog of Beeckman's library (four other books by Keckermann are).

48. *Journal*, 4:28, 31. So Van Laren denies matter having "duritas" (being absolutely hard).

49. Ibid., 4:30–31.

50. Ibid., 1:23.

51. According to the latest views, Hero was anything but a simple engineer who was interested in mechanical instruments without understanding the principles on which they were based. In fact, he had a position at the Museum at Alexandria comparable to a professorship. His *Pneumatica* probably is a collection of lecture notes on mechanical instruments.

See *Dictionary of Scientific Biography*, 6:310-15. On Hero's influence, see H. Schmitt, "Heron von Alexandria im 17. Jahrhundert," *Abhandlungen zur Geschichte der Mathematik* 8 (1898): 195-214; M. Boas, "Hero's 'Pneumatica': A Study of Its Transmission and Influence," *Isis* 40 (1949): 38-48.

52. Hero explicitly refers to the small vacuum, not to the larger vacuum. The large vacuum is the free space in which particles of matter can move freely; the small or interstitial vacuum is the free space that remains between contiguous round particles, small cavities in otherwise completely filled space. With Hero, Beeckman initially accepted only the interstitial vacuum, the "vacuum intermixtum" (*Journal*, 4:30). He later accepted the reality of the larger vacuum.

53. Ibid., 4:19.

54. Ibid., 1:72, 74-77.

55. Hieronymus Cardanus, *Opera omnia*, 10 vols. (Lyon: J. A. Huguetan and M.-A. Ravaud, 1663; reprint, Stuttgart: Fromman, 1966), 3:359, 365.

56. *Journal*, 1:278-80. Ctesibius was another ancient engineer Cardano introduced to Beeckman.

57. Ibid., 1:36. According to Beeckman, magnetism was caused by a flow of small particles (spiritus) from the magnet to and through the piece of iron, which was attracted to the magnet. "It is not without reason to say that this [attractive] spirit consists of barbed hooks, as Lucretius tells us" (Nec abs re hic dici potest spiritum hunc hamati figuris constare, ut loquitur Lucretius). For a more detailed analysis of Beeckman's notes on Lucretius, see Benedino Gemelli, *Isaac Beeckman. Atomista e lettore critico di Lucrezio* (Florence: Leo S. Olschki, 2002). See also chapter 6 in this volume.

58. *Journal*, 1:148.

59. Ibid., 1:201: "The difference in the essence of things depends on the disposition of the atoms" (Essentialis rerum differentia pendet ab atomorum situ). On the history of atomism, see Kurd Lasswitz, *Geschichte der Atomistik vom Mittelalter bis Newton*, 2 vols. (Hamburg: Leopold Voss, 1890; reprint, Hildesheim: Olms, 1963); M. Boas, "The Establishment of the Mechanical Philosophy," *Osiris* 10 (1952): 412-541; Lüthy et al., *Medieval and Early Modern Matter Theories*; Andrew Pyle, *Atomism and Its Critics: From Democritus to Newton* (Bristol: Thoemmes, 1995); Antonio Clericuzio, *Elements, Principles, and Corpuscles: A Study of Atomism and Chemistry in the Seventeenth Century* (Dordrecht: Reidel, 2000). Beeckman could draw on various authors for information about ancient atomism: Hero, Lucretius, Galen, Aristotle (in the *Journal* there is no reference to his *De generatione et corruptione*, but there are references to his *Meteorology*) and, among the moderns, Fernel and Basson.

60. *Journal*, 1:152-53; 3:138: "Atomi videntur tantum esse quatuor generum."

61. H. H. Kubbinga, *L'histoire du concept de "molecule,"* 3 vols. (Paris: Springer, 2002), 1:215, pictures the four basic atoms as a rhomb, a rectangle, a lens, and a hexagon (of different sizes). Although this is helpful in a way, Kubbinga acknowledges that these forms are a product of his own imagination; Beeckman never speculated about the precise form of atoms.

62. *Journal*, 1:152-53, 201-2; 2:85-86, 138, 280.

63. Ibid., 2:98, 103.

64. Bruce T. Moran, *Distilling Knowledge: Alchemy, Chemistry, and the Scientific Revolution* (Cambridge, MA: Harvard University Press, 2005), 98.

65. On the importance of illustrations (and art in general) for early modern science, see William B. Ashworth Jr., "Iconography of a New Physics," *History and Technology* 4 (1987):

267–97; Ashworth, "The Scientific Revolution: The Problem of Visual Authority," in *Conference on Critical Problems and Research Frontiers in History of Science and History of Technology* (Madison: University of Wisconsin Press, 1991), 326–48; Brian S. Baigrie, *Picturing Knowledge: Historical and Philosophical Problems concerning the Use of Art in Science* (Toronto: University of Toronto Press, 1996); Wolfgang Lefèvre, Jürgen Renn, and Urs Schoepflin, eds., *The Power of Images in Early Modern Science* (Basel: Birkhäuser, 2003); Christoph Lüthy, "Where Logical Necessity Becomes Visual Persuasion. Descartes' Clear and Distinct Illustrations," in Sachiko Kusukawa and Ian Maclean, eds., *Transmitting Knowledge: Words, Images and Instruments in Early Modern Europe* (Oxford: Oxford University Press, 2006), 97–133.

66. *Journal*, 1:211.

67. Before Beeckman, only Giordano Bruno had drawn pictures of atoms. The sources for Beeckman's realistic representation of the atom will be discussed in chapter 6.

68. *Journal*, 1:153: "It can happen that two substances have the same number of particles of fire, air, water, and earth, and still are different. In one case the particle of fire is in between earth and air, in another between air and water, and in general there are many differences in disposition of the four elements either placed in a line or arranged in a cube. In a plane these four elements give four figurations; in a square sixteen and in a cube 64" (Fieri enim potest ut duae res aequalibus constent portionibus corporum ignis, aeris, aquae et terrae, suntque tamen dissimilis naturae. Nam hisce sita est ignis particula inter terram et aerem, et etiam inter aeram et aquam, omninoque multae sunt quatuor simplicium figurarum in una linea dispositarum aut in forma cubi redactarum, positurae diversitates. In plano igitur haec quatuor elementa sunt quatuor hae figurae; ad formam quadrati disposita, dant sexdecim differentias, ad formam vero cubi 64). See Kubbinga, *L'histoire*, 1:203–25.

69. *Journal*, 2:57, 122, 127. For Beeckman's indebtedness to Lucretius, see Gemelli, *Beeckman*, esp. 57–59.

70. *Journal*, 1:163–64; 2:56.

71. Ibid., 1:23: "Nimirum ut architectus parat primordia domus: januam, fenestram, postes, trabes, tegmen, lapides, utque Rex Salomon singula primordia templi ita concinnavit ut absque pulsu mallei partes coirent sibique invicem quam ornatissime responderent, sic Deum naturae naturalia primordia fabricasse, quae sibi invicem ita conveniunt, porique clavis respondent, ut definitae res inde oriantur: lapides, arbores, animalia."

72. Ibid., 2:57.

73. Ibid.: ". . . ideoque minorum bestiarum creationem huic concursui permisit." Beeckman believed in the spontaneous creation of smaller animals, like worms emerging from cheese and flies emerging from manure.

74. Ibid., 2:124–25: "Si igitur primordia nostra forent tales pyramides ordinatae, et ad constitutionem speciei virtutes activas exerentes, requiretur compositum ordinatum, circulo inscribendum. Possent duntaxat duae esse species diversae: una quae constaret ex octo pyramidibus apte junctis, altera ex viginti; nam sex, septem etc. triangula conjuncta, constituerent quidem figuram aliquam, at non talem quae principium actionis in se contineret, juxta hypothesin. Constituant igitur icosahedra, apte sibi invicem conjuncta, hominem vel hominis semen; octahedra vero canem. . . . Primordia igitur physica habent quandam figuram, ex qua possunt quaedam minima componi, eaque finita, quae principium sint rationis, sensus, vitae, virtutis etc." Beeckman did not say exactly how these mental attributes emerged from the right ordering of the material primordia. One is tempted to conclude that

he too had to introduce active principles in order to explain the origin of life, the more so because he also speaks about the "semen" of man and of a semi-independent "vis agendi." See ibid., 2:57: "Once God created the principles of things, he left to them the force to act" (Deus semel rerum principia ponens, vim agendi iis reliquit). Yet, in fact, he only says that these active principles emerge from a specific ordering of inert primordial material.

75. Ibid., 2:118–19, 120–22, 127–28, 386. According to Kubbinga, Beeckman took the term "homogeneum" from Euclid's *Elements*, 5.3. Kubbinga, *L'histoire*, 1:217–19. He also refers to Viète's use of the concept of the "lex homogeneorum." De Wreede, on the other hand, quotes a passage of Willebrord Snellius's *Apollonius batavus* (1608), where Snellius states that mathematics, just like all sciences, comprises "many, but homogeneous topics" (plurima, eaque homogenea) in its sphere. See Liesbeth de Wreede, *Willebrord Snellius (1580–1626): A Humanist Reshaping the Mathematical Sciences* (Utrecht: printed by the author, 2007), 62–63. In a note, the author adds that the word homogeneous is probably "another Ramist reference," but she does not elaborate on this remark. Beeckman used the word homogeneum first in a pharmaceutical context, namely in a discussion of a statement made by Galen in *De theriaca ad Pisonem liber* (*Journal*, 2:118).

76. Beeckman is rather confusing in his designation of his particles of the second order. Sometimes it looks as though a homogeneum consists of primordia, while primordia in turn consist of atoms. Beeckman, however, was not really interested in establishing the actual constitution of matter; he merely wanted to show that an explanation of natural phenomena along these lines was possible and therefore that reliance on occult powers or inherent qualities was unnecessary. In this respect, Beeckman resembled Lucretius, who advanced his atomism because he wanted to show that one should not fear the gods.

77. *Journal*, 2:100–101.

78. In this respect, Beeckman was quite exceptional. Familiarity with chemical processes usually forced natural philosophers to accept the necessity of introducing active principles in their explanation of natural phenomena. The case of Daniel Sennert is particularly instructive. See Emily Michael, "Sennert's Sea Change: Atoms and Causes," in Lüthy et al., *Medieval and Early Modern Matter Theories*, 331–62, esp. 357–59.

79. *Journal*, 1:278–80; 2:254.

80. Ibid., 1:46; 4:30.

81. Ibid. 2:105–6, 157, 230.

82. Ibid., 3:224.

83. One may conclude that Beeckman was not able to find a reasonable explanation for the origin of life. To answer the question "quomodo ex insensibilibus fiat sensile" (an almost direct quote from Lucretius's *De rerum natura*, 2.888; cf. note 26 in this chapter), Beeckman could refer only to the atomist theory of Epicurus, as presented by Lucretius. Beeckman to Mersenne, April 30, 1630 (*Journal*, 4:186). See Gemelli, *Beeckman*, 11–15.

84. Beeckman mentions the "fuga vacui" for the first time in 1613/14 (*Journal*, 1:26). On this aspect of Beeckman's work, see Cornelis de Waard, *L'expérience barométrique. Ses antécédents et ses explications* (Thouars (Deux Sèvres): Imprimerie nouvelle, 1936), 75–93, 145–68; Giancarlo Nonnoi, *Il pelago d'aria. Galileo, Baliani, Beeckman* (Rome: Bulzoni Editore, 1988). On the concept of fuga vacui, see Boas, "Establishment of the Mechanical Philosophy," 117; Grant, *Much Ado about Nothing.*

85. *Journal*, 4:44: "Aqua suctu sublata non attrahitur vi vacui, sed ab aere incumbente in locum vacuum impellitur."

86. Ibid., 1:200.

87. Ibid., 1:46; 2:157.

88. Ibid., 1:135.

89. Ibid., 1:26, 36.

90. Ibid., 1:102, 2:49; 3:214.

91. Ibid., 1:81–82. For comparable explanations of the cohesion of two flat iron plates and the sphere-like shape of a drop of water, see ibid., 1:281; 3:177.

92. See John Henry, "Occult Qualities and the Experimental Philosophy: Active Principles in Pre-Newtonian Matter Theory," *History of Science* 24 (1986): 335–81; Osler, "How Mechanical Was the Mechanical Philosophy?"; Clericuzio, *Elements, Principles, and Corpuscles*.

93. *Journal*, 2:339.

94. Ibid.

95. Ibid., 3:17.

96. Magnetism was a well-known phenomenon, but only Gilbert in his *De magnete* in 1600 clearly distinguished it from the cohesive effect of (natural) electricity. Beeckman first saw Gilbert's book in 1627 when he was able to borrow it from Colvius (*Journal*, 3:17). The notes he subsequently took indicate that he recognized his differences with Gilbert: "What he says in chapter 4 that what is flowing from the magnet is not corporeal, I do not agree. I think that the stars radiate corporeal spirits to the magnetic Earth" (Quod vero cap. 4 dicit effluxum a magnete non esse corporeum, mecum non sentit. Existimo enim stellas in Terram magnetem immittere spiritos corporeos) (ibid., 3:17); and "While however Gilbert says that the movement of the Earth is star-like, that is, caused by an admirable inborn magnetic force, such that the Sun never perishes while burning, is that not like ascribing to the Earth an intelligence? Which is unworthy of a philosopher. He does not know how something that is incorporeal and what is external, moves [something corporeal]" (Cum tandem Gilbertus dicat Terrae motum esse astraeum, id est per insitam vim magneticam admirabilem, ne perpetuo Solis ardore pereat, an non videtur Terrae intelligentiam ascribere? Quod philosopho indignum. Nescit videlicet quomodo id quod inest incorporeum [quod vocat] quodque extra est, moveat) (ibid., 3:18). The best-known explanation of magnetism was put forward by Descartes in his *Principia philosophiae* (1644).

97. *Journal*, 1:36.

98. Gemelli, *Beeckman*, 3.

99. Ibid., 1:36; 2:339. See also Gemelli, *Beeckman*.

100. For Stevin's theory of the tides, see E. J. Dijksterhuis, *Simon Stevin* (The Hague: Martinus Nijhoff, 1943), 162–66; Dijksterhuis, *Simon Stevin: Science in the Netherlands around 1600* (The Hague: Martinus Nijhoff, 1970), 80–82.

101. *Journal*, 1:6.

102. Ibid., 2:167.

103. Ibid., 2:229. See also ibid., 2:317–18, 386–88; 3:11, 38, 86. Cf. Gemelli, *Beeckman*, 5.

104. *Journal*, 3:171, 205. On Galileo's theory of the tides, see H. I. Brown, "Galileo, the Elements and the Tides," *Studies in History and Philosophy of Science* 7 (1976): 337–51; M. A. Finocchiaro, *Galileo and the Art of Reasoning: Rhetorical Foundations of Logic and Scientific Method* (Dordrecht: Reidel, 1980), 74–79.

105. *Journal*, II1:206.

106. It cannot be determined whether Beeckman, in 1630, was familiar with Bacon's criticism of Galileo's theory, as put forward in his *Novum organum* (1620). Bacon rejected

Galileo's theory because it was based on the Copernican world system, which he found unacceptable. See E. J. Aiton, "Galileo's Theory of the Tides," *Annals of Science* 10 (1954): 44-57, esp. 48-49. During a visit to Middelburg in the summer of 1623, Beeckman had already seen the *Novum organum*, but he had had only a short time to read it through (the book was not on sale in Middelburg or anywhere else). In his excerpts of the *Novum organum*, the theory of the tides is not mentioned (*Journal*, 2:250-55). Also Kepler in his *Mysterium cosmographicum* of 1595 criticized the explanation of the tides as a consequence of movements of the Earth, but he had in mind the theory put forward by the sixteenth century Italian anatomist and botanist Andrea Caesalpino (Aiton, "Galileo's Theory," 48). Although Beeckman studied the *Mysterium cosmographicum* in the fall of 1628, he did not mention Kepler's criticism of Caesalpino. See *Journal*, 3:99.

107. *Journal*, 3:26. This note dates from November 1627. Beeckman had read Gilbert on October 8, 1627. See also Gemelli, *Beeckman*, 4.

108. *Journal*, 3:126. He also wondered what would happen if one were to put gold or sulfur "in a closed space from which almost all the air is extracted by a siphon" (in loco undique clauso indeque extracto per siphonem omni fere aere).

109. Ibid., 3:126-27.

110. Ibid., 3:239.

111. Ibid., 3:26.

112. For a survey of ether theories (in which Beeckman's name does not appear), see G. N. Cantor, introduction to G. N. Cantor and M. J. S. Hodge, eds., *Conceptions of Ether: Studies in the History of Ether Theories, 1740-1900* (Cambridge: Cambridge University Press, 1981), 1-60, esp. 1-19.

113. *Journal*, 1:25.

114. Ibid., 1:25; 2:18.

115. Several times Beeckman returned to the subject of ether as the cause of gravity; see ibid., 1:28; 2:107, 119-20, 232, 340; 3:25, 226.

116. Ibid., 1:1, 24. For a discussion of Beeckman's Copernicanism, see Rienk Vermij, *The Calvinist Copernicans: The Reception of the New Astronomy in the Dutch Republic, 1575-1750* (Amsterdam: Edita, 2002), 113-17.

117. *Journal*, 1:103-4.

118. Ibid., 1:104.

119. Ibid., 2:233.

120. Ibid., 2:232; 3:25.

121. Ibid., 3:28, 30.

122. Ibid., 1:38-39; 2:105; 4:183.

123. Ibid., 3:31, 218.

124. "Quas vocant Optici species visibiles sunt corporea." Ibid., 1:201. See also ibid., 4:44.

125. Ibid., 3:104.

126. Ibid., 1:251.

127. Ibid., 1:92-93; 2:71.

128. Ibid., 1:25, 99; 2:253; 3:49.

129. Ibid., 3:112, 349; 4:225-26.

130. J. Mac Lean, "De kleurenleer van de aanhangers der corpusculairtheorie," *Scientiarum Historia* 12 (1970): 1-22, esp. 6-10, gives a survey of the notes on colors in the *Journal*.

131. *Journal*, 2:198-99.

132. Ibid., 1:289; 2:220, 319.

133. Ibid., 1:32; 2:198; 3:239.

134. Ibid., 1:134, 154.

135. Ibid., 1:154, 155, 165, 216, 276.

136. Beeckman's opinions on the freezing of water changed over time. In 1614 he argued that ice occupies less space than water (ibid., 1:21–22); a year later he argued that ice occupies *more* space than water, the explanation being that the "natural heat" has to force its ways through the freezing water and thereby pushes the particles of water from each other (ibid., 1:60).

137. Ibid., 1:124, 310.

138. Ibid., 1:187; 2:198; 3:125.

139. In one of the *quodlibeta* attached to his 1618 dissertation, the Copernican position is thrown up for discussion: "The sun moves and the earth is at rest, or the earth moves and the sun is at rest" (Sol movetur & terra quiescit, aut terra movetur & sol quiescit) (ibid., 4:44).

140. Ibid., 3:74, 103. After having seen what Kepler had written about the motion of the Earth, Beeckman felt the wish to do so himself. "I hope some time to write a completed work on this matter, with my meditations, which he [Kepler] will not see, attached" (Spero me aliquando, meis meditationibus, quas ille [Kepler] non videbit, adhibitis, absolutum opus de hac re scripturum). Beeckman wrote these lines right after having reestablished contact with Descartes.

141. Ibid., 3:165.

142. There was some discussion in those years whether the third movement (which actually resulted in the standstill of the Earth's axis) in fact could be called a movement. Stevin said it could not. See *The Principal Works of Simon Stevin*, 5 vols. (Amsterdam: C. V. Swets and Zeitlinger, 1955–65), 3:130–32.

143. *Journal*, 3:108: "Hoc modo igitur ostendi omnes tres Terrae motus perfici absque ulla insita vi fictitia, et ex motu corpusculorum ex Sole ejaculatorum sequi consequutione mathematica."

144. Ibid., 1:101, 194.

145. Ibid., 2:382. It is questionable to write, as De Waard did: "She [the spiritus] penetrates, he thinks, all bodies without exercising any pressure. He [Beeckman] thus identifies it with the vacuum that separates the atoms, like Bruno" ("Elle pénètre, pense-t-il, tous les corps sans exercer par elle-même aucune pression; il l'identifie donc avec le vide qui sépare les atomes, comme Bruno"; *Correspondance du P. Marin Mersenne religieux minime*, 17 vols. [Paris: Beauchesne, 1932–88], 2:118). This identification of the ether and the vacuum is based on one note in the *Journal*, 1:101: "The spiritus does not press on the object but passes through it without any obstacle" ("Spiritus vero ille non incumbit rei: sed absque obstaculo transit"). However, Beeckman only intended to say that the form of the fire particles is such that they exactly match the pores in the iron, which explains why they do not exert pressure on the iron. From other notes, it is clear that in other instances the fire particles can indeed exert pressure, which is how gravity arises.

146. Nicolas Cusanus and Kepler also made the comparison between the daily rotation of the Earth and a spinning top (E. J. Aiton, *Vortex Theory of Planetary Motions* [London: Macdonald, 1972], 52). Descartes too uses the analogy in his *Principia philosophiae*. According to Descartes, the planets have been spinning around since their creation. Although such a motion is in principle finite, a decrease in such movement cannot be detected because the Earth

exists for only about five thousand or six thousand years. For a large object like a planet, such a span of time is comparable to a minute in the case of a small object like a spinning top. (In the slightly more extended French edition: "pour que, d'autant qu'un corps est plus grand, d'autant il peut retenir plus longtemps l'agitation qui luy a esté ainsi imprimée, et que la durée de cinq ou six mil ans qu'il y a que le monde est, si on la compare avec la grosseur d'une Planete, n'est pas tant qu'une minute comparée avec la petitesse d'une pirouette" (*Oeuvres de Descartes*, 9-2:193). In Latin, the word for spinning top (pirouette) was *turbo*, which should be translated as "tourbie" (ibid., 8-1:194). Compare also Descartes's explanation in his conversation with Burman in 1648 (ibid., 5:173), where he relates that he once had seen a top spinning for more than fifteen minutes.

147. *Journal*, 3:118–20, 142–43.

148. The same tension between a mathematical description and a corpuscular explanation can be found in Descartes's work. Descartes's corpuscular explanation, according to which the vortex of matter is the cause of the spinning movement of the planets, appears alongside his explanation of the movement of the planets using the mechanics of spinning tops. See *Oeuvres de Descartes*, 8-1:93–94. See also chapter 7.

CHAPTER 5: PRINCIPLES OF MECHANICAL PHILOSOPHY II: MOTION

1. On the Aristotelian theory of motion, see E. J. Dijksterhuis, *The Mechanization of the World Picture* (Oxford: Oxford University Press, 1961), 17–42; Alexandre Koyré, *Études galiléennes* (Paris: Hermann, 1939; reprint, 1966), 17–23; Koyré, "Galileo and Plato," in Koyré, *Metaphysics and Measurement: Essays in the Scientific Revolution* (London: Chapman & Hall, 1968), 16–43. An extensive treatment of Beeckman's opinions can be found in Alan Gabbey, "Descartes' Dynamical Thought: A Critical Study of Some Problems, Principles and Concepts" (Ph.D. dissertation, Queen's University Belfast, 1964). Gabbey follows Koyré's interpretations but with greater attention to detail. Peter Damerow et al., *Exploring the Limits of Preclassical Mechanics: A Study of Conceptual Development in Early Modern Science; Free Fall and Compounded Motion in the Work of Descartes, Galileo, and Beeckman*, 2nd ed. (New York: Springer-Verlag, 2004 [1991]), chap. 1: "Concept and Inference: Descartes and Beeckman on the Fall of Bodies," 9–69 (by Peter Damerow and Gideon Freudenthal), is highly critical of Koyré's point of view; see also: Richard Arthur, "Beeckman, Descartes and the Force of Motion," *Journal of the History of Philosophy* 45 (2007): 1–28.

2. Beeckman never cited Aristotle's *Physics*, books V–VIII, where Aristotle sets forth his theory of motion, but he must have been familiar with this theory through other books.

3. *Journal*, 1:24–25: "Quod vero Philosophi dicunt vim lapidi imprimi, absque ratione videtur. Quis enim mente potest concipere quid sit illa, aut quomodo lapidem in motu contineat, quave in parte lapidis sedem figat? Facillime autem mente quis concipiat, in vacuo motum nunquam quiescere, quia nulla causa mutans motum occurrit: nihil enim mutatur absque aliqua causa mutationis." This is not the first time Beeckman mentions his principle of inertia. We can also find it at the top of the same page of the manuscript: "Omnis res, semel mota, nunquam quiescit nisi propter externum impedimentum" (A thing, once moved, never comes to rest unless impeded by some external cause) (ibid., 1:24).

4. Ibid., 4:44: "Lapis ex manu emissus pergit moveri non propter vim aliquam ipsi accedentem, nec ob fugam vacui, sed quia non potest non perseverare in eo motu, quo in ipsa manu existens movebatur."

5. Ibid., 1:44. In Dutch: "Dat eens roert, roert altyt, soot niet belet en wort."

6. Ibid., 1:266. Somewhat later, however, Beeckman seems to use the term *impetus* as equivalent to "cause of motion." See ibid., 1:330: "Si navis velum remittat soloque priori impetu feratur, potest ita tamen clavo regi ut, facto motus sui semicirculo versus eandam, unde venit plagam, moveatur per eam lineam, qua venerat [et] uno eodemque impetu redeat" (When a ship strikes its sail, and continues to move solely because of its previously acquired impetus, it can be directed by the helm in such a way that it returns in the direction whence it came, with the same impetus). That the use of the word *impetus* is not a slip of the pen is demonstrated by the fact that around 1629 Beeckman added the marginal heading: "Impetus idem interdum navem in prioris contrariam plagam movet" (The same impetus sometimes moves a ship in a direction contrary to its prior direction).

7. Arthur, "Beeckman, Descartes and the Force of Motion," 11–12. According to Arthur, Beeckman held that motion and its concomitant force are both created *and preserved* by God. Yet Beeckman is not explicit about this. He usually leaves the impression that God, after creating the world and setting the objects in motion, left the world to itself (see, for instance, *Journal*, 1:132). Descartes is much more explicit on God preserving motion.

8. Gabbey demonstrated that in fact the term *perpetuum mobile* could mean different things: (1) an idealized system or device that indefinitely maintains the motion it already has, in the absence of dissipative influences; (2) a mechanical system that moves perpetually, overcoming dissipative influences and, in addition, sometimes doing some kind of useful work; (3) perpetually moving systems with internal physical sources of power, like magnetism and chemical combustion; and (4) a perpetually moving system using external sources of power, like the sun or atmospheric changes. He found that all proponents of the new natural philosophy of the seventeenth century denied the possibility of an idealized or a mechanical perpetuum mobile (1 and 2), but also that many of them allowed for the possibility of a magnetic, chemical, or meteorological perpetuum mobile (3 and 4). Alan Gabbey, "The Mechanical Philosophy and Its Problems: Mechanical Explanations, Impenetrability, and Perpetual Motion," in J. C. Pitt, ed., *Change and Progress in Modern Science* (Dordrecht: Reidel, 1985), 9–84, esp. 42–67.

9. *Journal*, 1:265: "In vacuo autem nullus est recessus a centro" (In a vacuum, there is no movement away from the center). Cf. ibid., 2:45: "Scripsi antehac corpora graviora in vacuo, ad nullam plagam inclinantia" (I have written before about heavy objects in a vacuum having no tendency to move in a particular direction).

10. Ibid., 3:180; Gabbey, "Descartes' Dynamical Thought," 13–18. For more about Descartes's influence, see my comments in chapter 7.

11. Beeckman to Mersenne, April 30, 1630 (*Journal*, 4:186): "Corporibus quies non est magis naturalis quam motus" (Rest is for objects no less natural than motion).

12. Ibid., 1:253: "*Id, quod semel movetur, in vacuo semper movetur*, sive secundum lineam rectam seu circularem, tam super centro suo, qualis est motus diurnus Terrae, [quam circa centrum, qualis est motus] annuus. Cum enim quaelibet minima pars circumferentiae sit curva, atque eodem modo curva atque tota peripheria, nulla ratio est cur motus circularis Terrae annuus desereret hanc lineam curvam et ad rectam procederet, nam recta non magis naturalis et aequalis naturae et extensionis est quam circularis, quia pars circumferentiae se eo modo habet ad totam, quo pars rectae ad rectam totam" (emphasis added by De Waard). The word *extension* refers to the medieval doctrine of the "configuration of qualities and motions." In a diagram the baseline or extension represents the subject (either the spatial extent of the substance or the time during which the quality is present); the lines perpendicular

to the baseline (latitude, intension, or gradus) represent the intensity of the quality. Damerow and Freudenthal, "Concept and Inference," 17–21.

13. Koyré denied that Beeckman held the modern concept of inertia (Koyré, *Études galiléennes*, 108–9, n. 3). Descartes's and Newton's notion of inertia indeed excludes circular movements. If an object remains in a circular movement, like a planet around the sun, this has to be explained by invoking a certain force. Gabbey agrees with Koyré and criticizes De Waard for ascribing a modern notion of inertia to Beeckman (Gabbey, "Descartes' Dynamical Thought," 18, 20, 22). He does not wish to go further than to ascribe to Beeckman an inertia "dictum." This is both excessively purist and finalistic. Beeckman may not have had "the" modern principle of inertia, but he certainly had a notion of inertia. The fact remains that Beeckman was one of the first to realize that only a change in motion needs an explanation. He did not realize that circular motion is an instance of a motion that is constantly changing. Gabbey has another argument against attributing the idea of inertia to Beeckman: it would be misleading to do so, because in the early seventeenth century inertia was still conceived of in the way Kepler had understood it, as a force, residing in the object, that resists motion and brings the object to rest. Claude Mydorge called this "a natural tardiness" (une tardivité naturelle) (Alan Gabbey, "Force and Inertia in the Seventeenth Century: Descartes and Newton," in Stephen Gaukroger, ed., *Descartes: Philosophy, Mathematics and Physics* [Brighton: Harvester Press, 1980], 230–320, esp. 288). Gabbey certainly has a point here, but common practice among historians of science warrants the ascription of a principle of inertia to Beeckman (and to Descartes as well).

14. "Censendum videtur coelum nec ab intelligentiis moveri, nec continuo Dei nutu, sed sua et situs natura semel motum, nunquam per se posse quiescere. *Quod ergo fieri potest per pauca, male dicitur fieri per plura*" (*Journal*, 1:10; emphasis added by De Waard). This note and the following, also a comment on Scaliger, were written shortly before Beeckman went to Saumur in 1612. Scaliger's book is not in Beeckman's auction catalog, so presumably he borrowed it from someone.

15. Galileo applied his principle of inertia only to circular movements, as in the example of an object moving without friction on the curved surface of the Earth.

16. *Journal*, 3:18.

17. As discussed in chapter 4, Beeckman later on found this "explanation" insufficient. He then devised a corpuscular mechanism to account for the continuous movement of the planets.

18. In modern notation, and in its most simple form, the law of free fall is expressed as $y = \frac{1}{2}gt^2$, y indicating the space traveled, t the time elapsed since the beginning of the fall, and g the constant force of attraction that was unknown at the time of Beeckman. See Dijksterhuis, *Mechanization*, 329–33; Koyré, *Études galiléennes*, 107–27; Gabbey, "Descartes' Dynamical Thought," 215–40; H. Watanabe, "On the Divergence of the Concept of Motion in the Collaboration of Beeckman and Descartes," in *Proceeding of the XIVth International Congress of the History of Science. Tokyo and Kyoto, 1974* (Tokyo, 1975), 2:338–40; Stillman Drake, "Free Fall from Albert of Saxony to Honoré Fabri," *Studies in History and Philosophy of Science* 5 (1975): 347–66; John Schuster, *Descartes and the Scientific Revolution, 1618–1634: An Interpretation* (Ann Arbor: University Microfilms International, 1977), 72–93; Damerow and Freudenthal, "Concept and Inference," 9–69; Arthur, "Beeckman, Descartes and the Force of Motion."

19. Damerow and Freudenthal, "Concept and Inference," 9.

20. Cornelis de Waard, "Eene correspondentie van Descartes uit de jaren 1618 en 1619," *Nieuw Archief voor Wiskunde*, Tweede Reeks, 7 (1905): 69–87.

21. Koyré, *Etudes galiléennes*; Koyré, *From the Closed World to the Infinite Universe* (Baltimore: Johns Hopkins University Press, 1957; reprint, 1968).

22. *Journal*, 1:44 (in Dutch). It is difficult to give a satisfactory translation of entries like these without using the concepts of classical mechanics, which acquired their meaning only *after* and as a *result* of the intellectual revolution of the early seventeenth century. Here, Beeckman assumes that there is a natural minimum of motion and that the actual motion of a falling object is the sum of a number of these minimum motions. This at least is consistent with his later remarks on the problem of free fall.

23. Damerow and Freudenthal, "Concept and Inference," 23–24.

24. *Journal*, 1:174: "Ratio est quia duo motus conjunguntur: primo naturalis deorsum tendens; secundo lapis semel motus in eo motu permanet, huicque motui denuo additur naturalis."

25. Ibid., 1:260–61. Translation by Damerow and Freudenthal (with slight corrections), as are the following quotes.

26. Ibid., 1:262.

27. Ibid.

28. Ibid., 1:263. It should be noted that Beeckman is not asking for a generalized expression of the relation between space and time; he is only interested in calculating specific values (one hour, two hours). Strictly speaking what we have here therefore is not "the" law of free fall. Beeckman poses his question as a craftsman would do; Descartes's answer is more that of a scholar. See Pamela H. Smith, "In a Sixteenth-Century Goldsmith's Workshop," in Lissa Roberts et al., eds., *The Mindful Hand: Inquiry and Invention from the Late Renaissance to Early Industrialization* (Amsterdam: Koninklijke Nederlandse Akademie van Wetenschappen, 2007), 43–44, where she discusses how artisans can indeed solve abstract mathematical problems in an "embedded" context.

29. *Journal*, 1:264. See Damerow and Freudenthal, "Concept and Inference," 38–39. For the word *hurtkens* (Italian *urti*, Latin *plagis*), see Benedino Gemelli, *Isaac Beeckman. Atomista e lettore critico di Lucrezio* (Florence: Leo S. Olschki, 2002), 2–3. Cf. *Journal*, 1:263: "Si igitur descensus lapidis fiat per distincta intervalla, trahente Terra per corporeos spiritus, erunt tamen haec intervalla seu momenta tam exigua, ut proportio eorum arithmetica, ob multitudinem particularum, non sensibiliter fuerit minor quam 1 ad 4. Retinenda ergo triangularis dicta demonstratio" (If, therefore, the fall happens through distinct intervals, because the Earth is attracting [the object] through corporeal spirits, these intervals or moments are very small, with the result that this arithmetical proportion, because of the multitude of particles, would not sensibly be less than 1 to 4. We, therefore, have to retain the said demonstration with triangles).

30. Arthur, "Beeckman, Descartes and the Force of Motion," 20.

31. Damerow and Freudenthal, "Concept an Inference," 39.

32. Ibid., 40.

33. Cf. *Journal*, 3:348, where Beeckman emphatically repeats that nature is discontinuous: "Alle dynghen gaen by horten, ut videre est cum magna pondera tarde moventur per machinas. Hinc probari potest omnia ex non in infinitum divisibilibus constare" (All things proceed by jerks, as we can see that heavy objects are moved by a machine. From that we can conclude that everything consists of parts that are not infinitesimally divisible).

34. Ibid., 1:261, 264.

35. Ibid., 3:226. Perhaps Beeckman used the word *attraction* simply because Descartes was using it.

36. *Le Opere di Galileo Galilei*, ed. Antonio Favaro, Edizione nazionale, 20 vols. (Florence: Tipografia G. Barbèra, 1890–1909; reprint, Florence: Barbèra, 1968), 8:197–267, esp. 197–98, 210–13; Galileo Galilei, *Two New Sciences, including Centres of Gravity and Forces of Percussion*, trans. with introduction and notes by Stillman Drake (Madison: University of Wisconsin Press 1974), 153–216, esp. 153–55, 166–70. Galileo had formulated his law of falling bodies in 1604 in a letter to Paolo Sarpi. See *Le Opere di Galileo Galilei*, 10:115; cf. Koyré, *Études galiléennes*, 86–87. Beeckman and Descartes learned about Galileo's opinion when they read the *Dialogo*. Beeckman acquired a copy in 1634 and, on visiting Descartes that year, lent it to him for two days (*Journal*, 3:356; 4:224–25).

37. For Descartes's rendering of the law of falling bodies, see *Journal*, 1:361 (1618); 4:49–50 (1618), 166–67 (1629), 170–71 (1629).

38. Damerow and Freudenthal, "Concept and Inference," 69, conclude: "What has been represented [by Koyré] as a comedy of errors thus turns out to be a coherent long-term discussion of an internally consistent argument. Beeckman understood Descartes' argument very well although he did not accept it. He noted it correctly in his *Journal* leaving out what was not directly related to his question and commenting on it from his point of view. Descartes consistently stuck to his argument because for understandable reasons he noticed neither the implied internal contradiction nor the differences to Beeckman's note insofar as these differences were not attributable to the differing physical theories of mechanics." The internal contradiction to which the authors refer concerns an ambiguity in the concept of "minimum moment of motion," which could have both a temporal and a spatial interpretation (ibid., 27). On the other hand, Arthur, "Beeckman, Descartes and the Force of Motion," 21, denies that Beeckman and Descartes disagreed as far as their theory of motion was concerned. "Notwithstanding the difference between Descartes and Beeckman over whether acceleration is discrete or continuous, there appears to be next to no difference of opinion between the two concerning the force of a body's motion." Arthur then proceeds to formulate nine basic assumptions of a common ontology of motion. If there had been any difference, he contends, Descartes would have certainly said so in 1630, but the fact is that he did not. Instead, he tried to claim this ontology of motion as his own.

39. Gabbey, "Descartes' Dynamical Thought," 248–49, 271–76.

40. On musical theory in the seventeenth century, see S. Dostrovsky, *The Origins of Vibration Theory: The Scientific Revolution and the Nature of Music* (Ann Arbor: University Microfilms International, 1969); Dostrovsky, "Early Vibration Theory: Physics and Music in the Seventeenth Century," *Archive for the History of Exact Sciences* 14 (1975): 169–218; C. V. Palisca, "Scientific Empiricism in Musical Thought," in H. H. Rhys, ed., *Seventeenth-Century Science and the Arts* (Princeton: Princeton University Press, 1961), 91–137; C. A. Truesdell, "The Rational Mechanics of Flexible or Elastic Bodies, 1638–1788," in *Leonhardi Euleri Opera Omnia*, ed. C. A. Truesdell, 2nd ser., vol. 11, part 2 (Turici [Zürich]: Orell Füssli, 1960), 14–141, esp. 24–28; D. P. Walker, *Studies in Musical Science in the Late Renaissance* (London: Warburg Institute, 1978); H. Floris Cohen, *Quantifying Music: The Science of Music at the First Stage of the Scientific Revolution, 1580–1650* (Dordrecht: Reidel, 1984), 24–28; Frédéric de Buzon, "Science de la nature et théorie musicale chez Isaac Beeckman (1588–1637)," *Revue d'Histoire des Sciences* 38 (1985): 97–120. Cohen's book is especially important because he also dis-

cusses topics discussed by Beeckman that are left out of consideration here. See also Patrice Bailhache, "Isaac Beeckman a-t-il démontré la loi des cordes vibrantes selon laquelle la fréquence est inversement proportionelle à la longueur?" *Revue d'Histoire des Sciences* 45 (1992): 337–44.

41. Aristotle's, *De anima*, 420a30–32. Whereas Aristotle simply states that a higher tone is produced in less time than a lower tone, Beeckman gives an exact proof of the relationship between length and pitch. Descartes's 1630 scorn of Beeckman's accomplishment is therefore completely out of place. In trying to belittle Beeckman's achievements, Descartes claimed that if Beeckman had read a little further in Aristotle than he had done in teaching his pupils, he would have learned that Aristotle had already stated what he, Beeckman, believed he had discovered. Descartes to Beeckman, October 17, 1630 (*Journal*, 4:199). Descartes repeatedly portrayed Beeckman as a conceited schoolteacher. In a November 4, 1630, letter to Mersenne, Descartes wrote in the same vein: "Car si j'ecrivois jamais de la morale, et que je voulusse expliquer combien la sotte gloire d'un pedan est ridicule, je ne la sçaurois mieux representer qu'en y mettant ces quatre lettres" (Because if I ever would want to write about morals, and would like to make it clear how ridiculous the vainglory of the pedant actually is, I would not know how better to do this than by presenting these four letters [exchanged between Beeckman and Descartes]" (ibid., 4:202–3). But as the *Journal* demonstrates, Descartes had learned about the relation between length, pitch, and frequency from Beeckman before he elaborated on it in his *Compendium musicae*.

42. *Journal*, 1:54–55.

43. In 1620 Beeckman applied the concept of *locus pausae* to rays of light reflected by the surface of the Earth: "In contactu Terrae radii ante reditum ad coelum necessario quiescunt; sic enim omne corpus, velut pila a muro repulsa, in puncto contactus nonnullo temporis momento quiescet, nam quies semper est inter duos motus contrarios eiusdem corporis" (On contact with the Earth, the rays necessarily rest before returning to the sky. In this way all objects, such as a ball rebounding from the wall, rest for a moment at the point of contact, because there is always rest between two opposing movements of the same object) (*Journal*, 2:23).

44. Benedetti and Galileo denied the existence of such a point of rest in 1585 and 1592, respectively. *Correspondance du P. Marin Mersenne religieux minime*, 17 vols. (Paris: Beauchesne, 1932–88), 2:400–401.

45. *Journal*, 1:249, 259.

46. Ibid., 1:260.

47. Ibid., 3:184–85; cf. 174–75.

48. Ibid., 3:183–84, 192–94, 198, 341. Beeckman explicitly refers to Santorio Santorio's pulsilogium when he constructed a metronome for the city musicians, who practiced in the Latin school. Ibid., 3:183. Santorio was an Italian physiologist and professor of theoretical medicine at the University of Padua, who described an instrument to measure the pulse in his 1625 *Commentaria in primam Fen Avicennae*, which Beeckman saw for the first time in December 1630.

49. *Journal*, 3:51–52: "Ego vero (ut etiam hic vides) multo ante videor eam reperisse. Crediderim enim Verulamium in mathesi cum physica conjungenda non satis exercitatum fuisse; Simon Stevin vero meo juditio nimis addictus fuit mathematicae, ac rarius physicam ei adjunxit."

50. For a discussion of Stevin's musical work, see Cohen, *Quantifying Music*, 45–66.

51. E. J. Dijksterhuis, *Simon Stevin: Science in the Netherlands around 1600* (The Hague: Martinus Nijhoff, 1943), 273–76. On Bacon, see Dostrovsky, *Origins of Vibration Theory*, 121–31.

52. Beeckman to Mersenne, October 1, 1629 (*Journal*, 4:157); cf. ibid., 1:29, 88–89, 180–81.

53. Ibid., 1:28–29; 2:34, 301–2, 450.

54. Ibid., 1:92.

55. Ibid., 1:92–93, 252–55; 4:214–15.

56. Ibid., 1:252–53.

57. Beeckman attributed this phenomenon to a natural feeling for singing of the common people, ignored by those who wrote on the theory of music. In a September 10, 1628, note, he concluded "that the community, overcome by nature, corrects the printed mistakes, against the precentor" (ibid., 3:80; see also ibid., 3:35–36).

58. The attention Beeckman gave to the role of the medium also is demonstrated in his 1619 remarks about the mast experiment. Those who denied the daily rotation of the Earth always pointed out that if the Earth did execute this movement, objects falling from a certain height would not fall down vertically, but with a deviation toward the west (because, according to the Copernican theory, during the fall the Earth had moved to the east). Copernicans, however, maintained that the object participated in the eastward movement of the Earth and thus executed an apparently vertical movement. They also pointed to the case of the stone that is dropped from the mast of a ship that is moving in whatever direction. This stone, so they argued, would fall down exactly at the foot of the mast and not behind the ship. Gassendi presumably was the first to really do the experiment (Howard Jones, *Pierre Gassendi, 1592–1655: An Intellectual Biography* [Nieuwkoop: De Graaf, 1981], 61–63, 83–84). Because on a sailing ship the wind might be an interfering factor, Beeckman suggested that the experiment be performed on a horse-drawn barge in one of the Dutch canals: "navis . . . equis tracta per funem, sicut in Hollandia passim fit" (*Journal*, 1:331).

59. *Journal*, 1:167, 253–55; Gabbey, "Descartes' Dynamical Thought," 58–60.

60. *Journal*, 2:335–38; cf. ibid., 1:30–33, 242–43, 362–63.

61. Ibid., 1:174.

62. Ibid., 1:150.

63. Ibid., 3:226.

64. Ibid., 4:124–25.

65. Although Beeckman does not mention it explicitly, he was probably familiar with the experiment executed by Stevin, in cooperation with the Delft burgomaster Jan Cornets de Groot and described in Stevin's *Anhang van de Weeghconst* (first published in 1586 and reprinted in 1605 in the *Wisconstighe Ghedachtenissen*). See *The Principal Works of Simon Stevin*, 5 vols. (Amsterdam: C. V. Swets and Zeitlinger, 1955–65), 1:510–11. Stevin had dropped objects from the leaning tower of the Old Church in Delft and demonstrated that objects do not fall faster when they are heavier. Beeckman, however, explained that still, because of the resistance of the air, there had to be some difference between the heavy and the light objects.

66. Beeckman to Mersenne, middle of March 1629 (*Journal*, 4:142).

67. Ibid., 1:267–68.

68. Ibid., 4:161, 184.

69. Thus far, historians of science have paid far more attention to Beeckman's notes on the law of falling bodies than to his notes on the laws of collision, although his laws of colli-

sion were far more important for his mechanical philosophy. Historians have compared Beeckman's rules with those of Descartes. Gabbey, "Descartes' Dynamical Thought," 88–104; Gabbey, review of W. L. Scott, *The Conflict between Atomism and Conservation Theory*, *Studies in History and Philosophy of Science* 3 (1972–73): 373–85; J. Mac Lean, *De historische ontwikkeling der stootwetten van Aristoteles tot Huygens* (Amsterdam: Vrije Universiteit, 1959), 16–19, 27–28; R. C. Taliaferro, *The Concept of Matter in Descartes and Leibniz* (Notre Dame: University of Notre Dame Press, 1964), 17–21 (with a systematic comparison, in modern notation, of the rules); Cornelis de Waard, "Sur les règles du choc des corps d'après Beeckman," in *Correspondance de Mersenne*, 2:632–44 (a survey of all the notes concerning collision in the *Journal*); Gabbey, "The Mechanical Philosophy and Its Problems"; Stephen Gaukroger and John Schuster, "The Hydrostatic Paradox and the Origins of Cartesian Dynamics," *Studies in the History and Philosophy of Science* 33 (2002): 535–72, esp. 555–56.

70. *Journal*, 1:265–66. Beeckman's comparison with statics is quite remarkable. He starts from a dynamic interpretation of the balance, and whereas Stevin in his *Wisconstighe Ghedachtenissen* (1605) had formulated a relation between force and distance (virtual displacements instead of virtual velocities), Beeckman looks for a relation between force and velocity. Apparently, he tried to lump together statics and collision theory. Cf. *Journal*, 3:133–34; Gabbey, "Descartes' Dynamical Thought," 89–90; Schuster, *Descartes*, 66–68. Cf. also Beeckman's attempts to determine the force of falling bodies by using a balance, an idea derived from Scaliger (*Journal*, 1:267; Mac Lean, *Historische ontwikkeling*, 9–11, 16–19).

71. *Journal*, 3:129.

72. Gabbey, "Descartes' Dynamical Thought," 95–96, 99–100.

73. *Journal*, 1:266.

74. De Waard already noticed this: ibid., 1:266, n. 1.

75. Ibid., 1:265, n. 6.

76. Ibid., 2:45.

77. Ibid., 1:25. Cf. 2:280.

78. Gabbey, "Descartes' Dynamical Thought," 101–3.

79. *Journal*, 1:266–67.

80. Gabbey, "Descartes' Dynamical Thought," 101–2. See also Gabbey, "The Mechanical Philosophy and Its Problems," 38–39.

81. See also Beeckman's remark from 1623: "Resultus aut reflectio laminarum, pilarum aeris, tam multis ante a me est exagitata ut fere desperaverim posse eius rei rationem reddi per atomos, id est corpusculis absque poris, minimis naturalibus existentibus" (I have puzzled so many times over the rebound and reflection of plates and balls filled with air that I despair of being able to explain these phenomena with the help of atoms, that is corpuscles without pores, being the smallest natural parts that exist) (*Journal*, 2:230).

82. Ibid., 3:128–29; Gabbey, "The Mechanical Philosophy and Its Problems," 39.

83. *Journal*, 3:369.

84. "Nihil igitur aut motus aut substantiae perit in mundo" (ibid., 2:139). It is not clear whether Beeckman believed that a certain amount of time could pass before an object could spring back to its original form. If he really toyed with the idea of "postponed motion," it seems that he was not far from the idea of potential energy.

85. Newton, *Opticks*, 1730 ed. (reprint, 1952), 398, cited in W. L. Scott, *The Conflict between Atomism and Conservation Theory, 1644–1860* (London: Macdonald, 1970; reprint, 1976), 4. On the ancients, see Gabbey, review of Scott, *The Conflict*.

86. Cohen, *Quantifying Music*, 159-60.

87. This is exactly why the traditional Aristotelian philosophy of nature persisted in schools and universities long after its basic principles had been discarded by most philosophers. Ann Blair, "Humanist Methods in Natural Philosophy: The Commonplace Book," *Journal of the History of Ideas* 53 (1992): 541-51, esp. 47-48. For another example of a commonplace natural philosophy, see Francis Bacon's *Sylva sylvarum* (1621), not surprisingly a book highly valued by Beeckman. On Beeckman as a reader of Bacon, see Benedino Gemelli, *Aspetti dell'atomismo classico nella filosofia di Francis Bacon e nel seicento* (Florence: Leo S. Olschki, 1996), chap. 5: "Isaac Beeckman e Bacone," 197-249.

88. The auction catalog of Beeckman's library mentions among the "Miscellanei in Octavo" the "Philosophia Nicolai Hill," but this undoubtedly refers to the 1619 reprint of the book, which may have been acquired by Beeckman much later. In the *Journal* there is no reference to Hill. Harriot is said to have done telescopic observations already in July 1609 because of his "long-standing personal contacts" with the Netherlands (John V. Pepper, "Harriot, Thomas," *New Dictionary of Scientific Biography*, 8 vols [Detroit: Charles Scribner's, 2008], 3:245-47), but Beeckman never mentions him in his notebooks. Harriot, the excellent mathematician, the Copernican astronomer, the atomist, and the co-discoverer of the sine law of refraction, was first and foremost a problem solver, whereas Beeckman was primarily a natural philosopher. Harriot corresponded with Kepler, but not much about the content of their letters seems to have filtered through to other mathematicians and natural philosophers.

89. Kurd Lasswitz already noted that around 1620 the corpuscular theory (which is not identical to the mechanical philosophy and can be combined with concepts that are utterly unmechanical) was very much "in the air," referring to Daniel Sennert, Hill, Bacon, Gorlaeus, Basson, Jean d'Espagnet, and Galileo. The crucial point, however, is that Beeckman never read their works or only did so (as with Basson) after devising his own system. On Gorlaeus, see Christoph Lüthy, "David Gorlaeus' Atomism," in Christoph Lüthy et al., *Medieval and Early Modern Matter Theories* (Leiden: Brill, 2001), 245-90, esp. 260, and also Lüthy, *David Gorlaeus (1591-1612): An Enigmatic Figure in the History of Philosophy and Science* (Amsterdam: Amsterdam University Press, 2012). Beeckman owned a copy of Gorlaeus's *Exercitationes philosophicae* (1620), bound together with Pieter van Foreest, *De Lue Venera* (*Observationum et curationum medicianalium, sive Medicinae Theoricae et practicae libri XXX, XXXI et XXXII. De Venenis, Fucis et Lue Venera: In quibus eorundem causae, signa, prognosis et curationes graphice depinguntur* [Leiden, 1606], reprinted several times). See Benedino Gemelli, "Il Catalogus Librorum di Isaac Beeckman," *Nouvelles de la République des Lettres*, 1991, no. 1 (1992): 131-59, esp. 146. Beeckman probably acquired Gorlaeus's book long after having devised his own corpuscular philosophy. At least, he never referred to it. He did, however, refer to book 19 of Foreest's *Observationes*, which deals with scurvy (*Journal*, 1:146).

CHAPTER 6: SOURCES FOR A MECHANICAL PHILOSOPHY

1. *Journal*, 2:84. His teacher was the mathematical practitioner Jan van den Broecke.

2. Ibid., 1:244.

3. Vermij provides some evidence for scientific contacts Lansbergen had in Middelburg before his move there from Goes in 1613. Lansbergen's first mathematical work, on trigonometry—published in Leiden in 1591—was dedicated to the city government of Middelburg. In the dedication, Lansbergen states that he had written the book "in vestra hac in-

sula" (meaning on the island of Walcheren); in the second edition of 1631, he writes: "primum in urbe vestri concepi [i.e., in Middelburg], post Goesae scripsi" (I first conceived of this in your town, and afterward wrote it down in Goes). Rienk Vermij, *The Calvinist Copernicans: The Reception of the New Astronomy in the Dutch Republic, 1575–1750* (Amsterdam: Edita, 2002), 75, note.

4. John Henry, "Occult Qualities and the Experimental Philosophy: Active Principles in Pre-Newtonian Matter Theory," *History of Science* 24 (1986): 335–81. The widely accepted view that it is possible to distinguish between a vitalistic and a mechanistic conception of matter in the early seventeenth century is now believed no longer to hold. Antonio Clericuzio, *Elements, Principles and Corpuscles: A Study of Atomism and Chemistry in the Seventeenth Century* (Dordrecht: Reidel, 2000); Bruce T. Moran, *Distilling Knowledge: Alchemy, Chemistry and the Scientific Revolution* (Cambridge, MA: Harvard University Press, 2005).

5. Stevin had settled in the North in 1580; Beeckman's father had come to Middelburg in 1588.

6. W. R. Shea, as quoted in Jim Bennett, "The Mechanic's Philosophy and the Mechanical Philosophy," *History of Science* 24 (1986): 1–28, quotation on 4. M. Boas, "The Establishment of the Mechanical Philosophy," *Osiris* 10 (1952): 412–541, also posits a direct link between ancient atomism and modern mechanistic philosophy.

7. On the popularity of Lucretius in the Low Countries, see Piet Schrijvers, "Schildknaap en tolk van Epicurus. Lucretius in Nederland," in Lucretius, *De natuur van de dingen (De rerum natura)*, uitgegeven, vertaald, ingeleid en van aantekeningen voorzien door Piet Schrijvers (Groningen: Historische Uitgeverij, 2008).

8. Benedino Gemelli, *Isaac Beeckman: Atomista e lettore critico di Lucrezio* (Rome: Leo S. Olschki, 2002).

9. *Journal*, 1:272–73. Cf. Gemelli, *Beeckman*, 43–45.

10. *Journal*, 1:36; "Nec abs re hic dici spiritum hunc hamatis figuris constare ut loquitur Lucretius" (Not without reason one can say that this spirit consists of hooked figures). See Lucretius, *De rerum natura*, 2.445, 6.1002–8, 6.1024–41.

11. Likewise, Beeckman only read Basson's textbook of atomistic natural philosophy after having arrived at his own conclusions.

12. *Journal*, 3:49.

13. Elsewhere in the *Centuria*, Abraham Beeckman deleted Lucretius's name altogether, for instance, in a note discussing the question why wind gets stronger along the way (*Journal*, 3:31–32). Was this because Abraham no longer considered the similarity with Lucretius's opinion to be an argument in favor of his brother's ideas? Or had Lucretius become an uncomfortable ally by 1644? See Yasmin Haskell, "A Dutch Lucretian" [a review of Gemelli, *Beeckman*], *Classical Review* 54 (2004): 549–50: "One is left wondering whether Beeckman would have discoursed so nonchalantly with Lucretius in print—or in poetry" (ibid., 550).

14. Beeckman also noted a manuscript under the title of "Conste der distillatien" in the list of Stevin's literary remains, which suggests a textbook for chemistry or alchemy (*Journal*, 2:291). Beeckman, however, adds "van ander schrift," which means "not written in Stevin's hand." The actual author is unknown.

15. *The Principal Works of Simon Stevin*, 5 vols. (Amsterdam: C. V. Swets and Zeitlinger, 1955–65), 1:292; 3:620. See E. J. Dijksterhuis, *Simon Stevin: Science in the Netherlands around 1600* (The Hague: Martinus Nijhoff, 1970); Klaas van Berkel, "Spiegheling en daet bij

Beeckman en Stevin," *Tijdschrift voor de Geschiedenis der Geneeskunde, Natuurwetenschappen, Wiskunde en Techniek* 2 (1979): 89–100. On Beeckman and Stevin, see also Jozef T. Devreese and Guido Vanden Berghe, *"Wonder en is gheen wonder": De geniale wereld van Simon Stevin, 1548–1620* (Leuven: Davidsfonds, 2003), 310–12.

16. *Journal*, 2:291–93.

17. Ibid., 2:236.

18. In Beeckman's words: "Iam vero solum contigui existentes, non minus afferunt ponderis fundo quam ante" (ibid.).

19. Dijksterhuis, *Stevin* (1970), 53.

20. Helen Hattab, "From Mechanics to Mechanism: The *Quaestiones mechanicae* and Descartes' Physics," in Peter R. Anstey and John Schuster, eds., *The Science of Nature in the Seventeenth Century* (Dordrecht: Reidel, 2005), 99–129.

21. Beeckman repeatedly referred to the *Quaestiones mechanicae*, but I cannot ascertain which of the handful of sixteenth-century editions he actually consulted. The auction catalog simply mentions "Problemata Aristotelis & aliorum," which probably refers to another book, under "Miscellanei in 12. & 16."

22. Bennett, "The Mechanic's Philosophy."

23. See chapter 5, note 28.

24. *Journal*, 1:159.

25. Ibid., 1:104; 2:161–62; 3:75. I note, however, that Beeckman makes these remarks in the course of his reading the works of Jean Fernel and Jean Riolan (or Riolanus) the Elder, two major sixteenth-century French defenders of Galen. Beeckman even explicitly says that Riolan makes the comparison, "quite adequately, in my opinion" (non inepta sane, meo quidem juditio). Ibid., 2:161. Thus, what appears to be an observation from everyday life turns out to be, as in the case of Lucretius, a quotation from a book.

26. Ibid., 1:24–25.

27. Ibid., 3:162.

28. John Schuster, *Descartes and the Scientific Revolution, 1618–1634: An Interpretation* (Ann Arbor: University Microfilms International), 59.

29. Ibid. See also Stephen Gaukroger and John Schuster, "The Hydrostatic Paradox and the Origins of Cartesian Dynamics," *Studies in the History and Philosophy of Science* 33 (2002): 535–72, esp. 553: "The *Journal* testifies to his [Beeckman's] private goal of reforming natural philosophy in the name of the values of mechanical intelligibility and utility."

30. The literature amply documents the artisanal background to new philosophies of nature in early modern Europe. Edgar Zilsel, *The Social Origins of Modern Science*, ed. Diederik Raven et al. (Dordrecht: Kluwer Academic Publishers, 2000 [1976]); Paolo Rossi, *Philosophy, Technology and the Arts in the Early Modern Era* (New York: Harper & Row, 1970 [1962]); A. Clegg, "Craftsmen and the Origin of Science," *Science and Society* 43 (1979): 182–201; Bennett, "The Mechanic's Philosophy." For more recent literature, see the bibliographical essay in this volume.

31. See especially Pamela H. Smith, *The Body of the Artisan: Art and Experience in the Scientific Revolution* (Chicago: University of Chicago Press, 2004).

32. Beeckman was not the only one who was dissatisfied with the lack of knowledge among artisans. When Prince Maurice wanted to learn something about perspective, he went to some of the best painters he knew. Their explanations, however, did not satisfy him, and he discovered that with these painters "the shortening of the lines and the adjustment of

the angles is done through estimation or at a guess" (de vercorting der linien en verandering der houcken uyter oogh of bij der gisse toeginck), whereas he wanted to learn how to use perspective "with knowledge of causes and their mathematical proofs" (met der kennis der oirsaken en sijn wisconstig bewijs). For that reason, he called upon Stevin, who perhaps never had held a paintbrush but knew how to explain perspective in mathematical terms. E. J. Dijksterhuis, "Simon Stevin," *Jaarboek van de Maatschappij der Nederlandse Letterkunde te Leiden* 1950–1951, 43–63, esp. 56. Sometimes there was more than simple ignorance involved. The introduction of the new science of motion in the seventeenth century met with obstruction from artillerymen, who, like most practitioners, did not want to give up their rules of thumb. A. Rupert Hall, *Ballistics in the Seventeenth Century: A Study of the Relations of Science and War* (Cambridge: Cambridge University Press, 1952); cf. Alexandre Koyré, "Galileo and Plato," in Koyré, *Metaphysics and Measurement: Essays in the Scientific Revolution* (London: Chapman & Hall, 1968), 16–43, esp. 17: "The new ballistics was not made by artificers and gunners, but against them."

33. *Journal*, 3:306.

34. In the decades after 1600, customers in the shipbuilding industry increasingly demanded a scale model before giving their final commission to build the actual ship. Karel Davids, *The Rise and Decline of Dutch Technological Leadership: Technology, Economy and Culture in the Netherlands, 1350–1800*, 2 vols. (Leiden: Brill, 2008), 432. The science of scaling down was therefore much in demand.

35. *Journal*, 2:256. This might be the first time in history that mention is made of the conscious application of rules of scale. And Beeckman's inaugural address of 1627 is the first systematic treatment of these rules.

36. Ibid., 2:455; 3:61.

37. Ibid., 3:215.

38. Ibid.: "Een instinctus van de nature . . . die Godt gegeven heeft tot bewys van het eeuwigh leven hierna."

39. Eugenio Canone, "Il *Catalogus librorum* di Isaac Beeckman," *Nouvelles de la Republique des Lettres* 1991, no. 1 (1992): 131–59. It is possible that some of the theological books originally had been part of his brother Jacob's library; Jacob was even more versed in theology than Isaac Beeckman, but the simple fact that Beeckman kept these books testifies to his interest in the matter.

40. Max Weber, *The Protestant Ethic and the Spirit of Capitalism* (New York: Scribner, 1958 [German ed., 1905]).

41. Robert K. Merton, *Science, Technology and Society in Seventeenth Century England* (New York: Fertig, 1970 [*Osiris* 1938]).

42. Reyer H. Hooykaas, "Science and Reformation," *Journal of World History* 3 (1956): 109–39; Hooykaas, *Religion and the Rise of Modern Science* (Edinburgh: Scottish Academic Press, 1972). For a more balanced view, see David C. Lindberg and Ronald L. Numbers, eds., *God and Nature: Historical Essays in the Encounter between Christianity and Science* (Berkeley: University of California Press, 1986); John Hedley Brooke, *Science and Religion: Some Historical Perspectives* (Cambridge: Cambridge University Press, 1991).

43. *Journal*, 2:358.

44. Peter Harrison, *The Fall of Man and the Foundations of Science* (Cambridge: Cambridge University Press, 2007).

45. *Journal*, 1:229.

46. William Perkins, *An Exposition of the Creede*, as quoted in Paul H. Kocher, *Science and Religion in Elizabethan England* (San Marino, CA: Huntington Library, 1958), 103. Perkins's book does not figure in the auction catalog of Beeckman's books.

47. Perkins (a fellow of Christ College, Cambridge) was not a middle-of-the-road Aristotelian philosopher but rather a devoted Ramist. See Donald K. McKim, "Ramism in William Perkins' Theology," *Sixteenth Century Journal* 16 (1985): 503–17; McKim, *Ramism in William Perkins' Theology* (New York: Lang, 1987). Perkins was still popular in the Dutch Republic in Beeckman's time. His Ramus-like definition of theology as "doctrina bene vivendi" (the doctrine to live well) accorded perfectly with the pietist circles in which Beeckman moved.

48. *Journal*, 1:229. Beeckman wrote this in October 1618, during the Twelve Years' Truce, and so he was not referring to a particular victory. Otherwise, he followed the developments on the battlefield very closely, judging by a number of sheets and booklets in his library relating military events, such as the siege and taking of 's Hertogenbosch in 1629.

49. Ibid., 2:63.

50. Ibid., 2:57. For an explanation of the concept of "principia," see chapter 4.

51. The phrase "non possunt non" recurs at the end of the note: "In man, such primordia are collected that cannot but produce this man" (In hominis igitur corpore talia primordia collecta sunt quae non possunt non talem hominem efficere). See also *Journal*, 2:43, where Beeckman refers to Lucretius, but also argues that to attribute to God only the creation of the principia, leaving it to these principia to bring forth the individual natural objects, is not impious at all. God created the atoms in such a way that "the callus of a young camel, which is only of use once the camel is domesticated by man, is already formed in the womb of the mother camel" (callus camelo in ventre natus est qui non nisi longo post tempore in usum venit, hominibus videlicet id animal subigentibus: nam iis atomis concurrentibus ex quibus camelus constat, id callus *non potuit non* nasci). See also ibid., 1:23, where Beeckman compares God to an architect and to King Solomon.

52. Ibid., 2:63. Actually, Beeckman does not use the phrase "laws of nature." Instead he refers to the more general term "modus": the divinely prepared particles are forced together "in such a way that they cannot but congregate in the right way" (eo modo, ut non possint non bene concurrere).

53. Ibid., 3:30–31.

54. Ibid., 3:34.

55. Ibid., 3:188.

56. Ibid., 2:375.

57. Ibid., 1:341, cf. 1:281–82.

58. Ibid., 2:242.

59. Ibid., 2:241–42.

60. Ibid., 2:243.

61. This did not preclude Beeckman from thinking that "avoiding the influence of the stars" was helpful in maintaining good health (ibid., 1:228).

62. Ibid., 2:203. Beeckman treated Drebbel's son-in-law, Johan Sibertus Kufler, in exactly the same way. Kufler visited Beeckman several times in Dordrecht and tried to sell the inventions of his father-in-law, who had died in 1633. According to Beeckman, Kufler presented these inventions as even more fantastic than Drebbel had. Ibid., 3:3, 300, 367. Beeckman knew Kufler, who in contrast to Drebbel, had some academic training (he had studied at the Calvinist Hohe Schule in the German town of Herborn) through Justinus van Asche, who

had met Kufler in Cologne. Beeckman's opinion about the glassblower Jacobus Bernhardi was even more devastating. This man, familiar with several alchemists, boasted he could cure a wound without touching it. He disclosed his method to Beeckman, but Beeckman was not impressed. He stated: "I write this down in order to demonstrate the idiocy of such nonsense" (Dit schryve ick hier om te toonen de sotticheyt van sulcke beuselen) (ibid., 2:201–2). See also chapter 3.

63. *Journal*, 2:388, 389; 3:5–6.

64. Ibid., 3:18.

65. Ibid., 3:34: "This is not adducing a cause, but obscuring it" (Hoc enim non est causam proferre, sed eam occultare).

66. Ibid., 2:254: "These folds cannot, to my judgment, be perceived by the human mind, and should therefore be expelled from the true philosophy" (Eius plicae . . . nequeunt, meo juditio, ab ingenio humano percipi, ideoque exterminandae a vera philosophia); cf. Bacon, *Novum Organum*, ii, 48. For an exposition of Bacon's theory of matter, see Clericuzio, *Elements, Principles and Corpuscles*, 78–80; Benedino Gemelli, *Aspetti dell'atomismo classico nella filosofia di Francis Bacon e nel Seicento* (Florence: Leo S. Olschki, 1996).

67. See note 51.

68. Theo Verbeek, *Descartes and the Dutch: Early Reactions to Cartesian Philosophy, 1637–1650* (Carbondale: Southern Illinois University Press, 1992).

69. Beeckman was quite ecumenical in his contacts with people of various creeds. He not only corresponded with members of the Catholic clergy, including Gassendi and Mersenne, but also provided lodging to them when they traveled in the Dutch Republic. For a detailed analysis of the theological background of Gorlaeus's atomism, see Christoph Lüthy, *David Gorlaeus (1591–1612): An Enigmatic Figure in the History of Philosophy and Science* (Amsterdam: Amsterdam University Press, 2012).

70. *Journal* 1:10: "Quod ergo fieri potest per pauca, male dicitur fieri per plura." Cf. ibid., 3:26.

71. Ibid., 1:24–25.

72. On the notion of intelligibility as an essential ingredient of natural philosophy, see Peter Dear, *The Intelligibility of Nature: How Science Makes Sense of the World* (Chicago: University of Chicago Press, 2006).

73. William of Ockham, *Summa totius logicae*, pars I, cap. 12. Ockham's razor is usually phrased as "Entia non sunt multiplicanda praeter necessitatem" (entities must not be multiplied beyond necessity), but these are not Ockham's own words. In fact, the principle was articulated this way by the Irish Franciscan John Ponce in 1639. See Brad S. Gregory, *The Unintended Reformation: How a Religious Revolution Secularized Society* (Cambridge, MA: Belknap Press of Harvard University Press, 2012), 402, n. 31.

74. On Ramus's life and work, see Charles Waddington, *Ramus: Sa vie, ses écrits et ses opinions* (Paris: Ch. Meyrueis et Cie, 1855); Walter J. Ong, *Ramus, Method, and the Decay of Dialogue: From the Art of Discours to the Art of Reason* (Cambridge, MA: Harvard University Press, 1958; reprinted, with a foreword by Adrian Johns, Chicago: University of Chicago Press, 2004); Reyer Hooykaas, *Humanisme, science et Réforme: Pierre de la Ramée (1515–1572)* (Leiden: Brill, 1958); J. V. Skalnik, *Ramus and Reform: University and Church at the End of the Renaissance* (Kirksville, MO: Truman State University Press, 2002); *Autour de Ramus: Le combat*. Études réunies et présentées par Kees Meerhoff et Jean-Claude Moisan, avec la colloboration de Michel Magnien (Paris: Honoré Champion, 2005). Because my interpretation

of Ramus in large measure depends on the work of Ong, it may be helpful to provide some background here. Walter Jackson Ong was born in 1912 and became a Jesuit. He taught English literature in St. Louis and wrote a great number of books on language and culture. Many of his ideas grew out of his fascination for the thesis of his friend and former colleague Marshall McLuhan (1911-80), who in a highly provocative way exposed the influence of communication media on human consciousness (the slogan "The medium is the message" is McLuhan's). According to McLuhan's theory, the invention of written language and especially the invention of the printing press caused a shift in human consciousness from the ear to the eye, which led to a loss of balance between the human senses. Ong asserted that Ramus is partly responsible for this deplorable development. Ong was also very concerned about the massification, the impersonality, and the loneliness of the individual in modern society. According to Ong, Ramus contributed significantly to this impoverishment of the human condition because he stressed the visual element in his logic and thus the impersonal method of communication. My appreciation of Ong's interpretation of Ramus does not infer my endorsement of his cultural criticism. On Ong, see Betty Rogers Youngkin, *The Contributions of Walter J. Ong to the Study of Rhetoric: History and Metaphor* (Lewiston, NY: Mellen University Press, 1995). For a much-needed critical reassessment of Ong's views, see Howard Hotson, *Commonplace Learning: Ramism and Its German Ramifications, 1543-1630* (Oxford: Oxford University Press, 2007).

75. On Ramus's ideas about mathematics, see Hooykaas, *Humanisme*, 75-90, and especially J. J. Verdonk, *Petrus Ramus en de wiskunde* (Assen: Van Gorcum, 1966).

76. Verdonk, *Petrus Ramus*, 341-70.

77. See also J.-C. Margolin, "L'enseignement des mathématiques en France (1540-70): Charles de Bovelles, Fine, Pelletier, Ramus, " in P. Sharratt, ed., *French Renaissance Studies, 1540-70: Humanism and the Encyclopedia* (Edinburgh: Edinburgh University Press, 1972), 109-55, esp. 152.

78. Verdonk, *Petrus Ramus*, 346.

79. As Ramus put it, ". . . ut mathematicis elementis regia via dispositis propositiones singulae suam veritatem non solum proponant, sed ordinis sui argumento praecipue demonstrent" (quoted in ibid., 348).

80. According to Ramus, there was an evident correspondence between natural reason (given to all men) and the order of things in the world. Hooykaas, *Humanisme*, 51, puts it this way: "The central idea of Ramism is that truth reveals itself in natural reason, in human intelligence, and at the same time in the natural order of the things in nature. This means that, provided that one follows his nature (the natural capacity of the intellect to apprehend truth), there is a precise correspondence between our conception of things and the things themselves" (L'idée centrale du ramisme c'est que la vérité se révèle dans la raison naturelle, dans l'intelligence humaine et, en même temps, dans l'ordre des choses naturelles dans l'univers. Cela signifie que, pourvu qu'on suive sa nature [la puissance naturelle de l'intellect à apercevoir la vérité], il y aura une correspondance précise entre nos conceptions des choses et les choses elles-mêmes). This idea is also at the heart of Beeckman's trust in the capacity of the human mind to grasp the way nature works.

81. For a general survey of sixteenth-century logic, see Wilhelm Risse, *Die Logik der Neuzeit*, 2 vols. (Stuttgart-Bad Canstadt: Friedrich Frommann, 1964-70), esp. vol. 1 (1500-1640). Also see W. S. Howell, *Logic and Rhetoric in England, 1500-1700* (Princeton, NJ: Princeton University Press, 1956).

82. Risse, *Logik der Neuzeit*, 1:153.

83. Hotson, *Commonplace Learning*, 40–41.

84. Ibid., 87 (quoting from S. H. M. Galama, *Het wijsgerig onderwijs aan de Hogeschool te Franeker 1585–1811* [Franeker: T. Wever, 1954], 90). Hachting's words were directly aimed at his staunchly Aristotelian colleague Johannes Maccovius. According to his auction catalog, Beeckman owned a copy of the *Collegium Logicum Hachtingii*, which must be identical with Hachting's *Collegium Logico-Practicum, Analyticam, in quo varia themata tum profane, tum sacra logice resolvuntur* (of which no copy is extant). Hachting studied at the University of Franeker from 1611 to 1616 and may have been a friend of Jacob Beeckman and Jeremias van Laren.

85. Hotson, *Commonplace Learning*, 107–8.

86. Compare the illustrations in Ong, *Ramus*, 202, 261, and the accompanying explanations on p. xviii.

87. Ibid., 82. Commenting on a book on logic by the Ramist philosopher Thomas Murner, Ong remarks: "The Ramist would prefer quasi-geometrical inspection of words in space to such allegorical insight into symbol" (ibid., 91).

88. Friedrich Beurhaus, rector of the Dortmund Gymnasium, to Freige, as cited in Hotson, *Commonplace Learning*, 27.

89. C. O. Bangs, *Arminius: A Study in the Dutch Reformation* (Nashville: Abingdon Press, 1971), 59, citing I. Breward's unpublished biography of William Perkins.

90. Owen Hannaway, *The Chemists and the Word: The Didactic Origins of Chemistry* (Baltimore: Johns Hopkins University Press, 1975), 132. Hannaway here refers to the influence of Ramus on the chemist Andreas Libavius.

91. Ong, *Ramus*, 295; Ong, *Ramus and Talon Inventory: A Short-Title Inventory of the Published Works of Peter Ramus (1515–1572) and of Omer Talon (ca. 1510–1562) in Their Original and in Their Variously Altered Forms* (Cambridge, MA: Harvard University Press, 1958).

92. In general, see Ong, *Ramus*, 295–318; Hooykaas, *Humanisme*, 113–20; Joseph S. Freedman, "The Diffusion of the Writings of Petrus Ramus in Central Europe, c. 1570–c. 1630," *Renaissance Quarterly* 46 (1993): 98–152; Mordechai Feingold, Joseph S. Freedman, and Wolfgang Rother, eds., *The Influence of Petrus Ramus: Studies in Sixteenth and Seventeenth Century Philosophy and Sciences* (Basel: Schwabe, 2001) (see also the review by Howard Hotson in *History of Universities* 19 [2004]: 192–96); Hotson, *Commonplace Learning*.

93. On Snellius, see Johannes Meursius, *Atheneae Batavae, sive de urbe Leidensi ac Academia, virisque claris . . . libri duo* (Leiden: Andreas Clouck et al., 1625), 117–22; *Nieuw Nederlandsch Biografisch Woordenboek*, vol. 7, cols. 1152–55; Liesbeth de Wreede, *Willebrord Snellius (1580–1626): A Humanist Reshaping the Mathematical Sciences* (Utrecht: printed by the author, 2007), 36–46. On Ramism in the Netherlands, see Paul Dibon, "L'influence de Ramus aux universités néerlandaises du 17e siècle," in *Proceedings of the XIth International Congress of Philosophy, Brussels, 1953* (Brussels, 1953), 14:307–11; Theo Verbeek, "Notes on Ramism in the Netherlands," in Feingold, Freedman, and Rother, *The Influence of Petrus Ramus*, 38–53. Given that Beeckman was born and raised in Middelburg, it is perhaps interesting to note that a preacher of the English Church in Middelburg, Dudley Fenner (1558–87), published *The Artes of Logike and Rhetorike*, which essentially is an English translation of Ramus's *Dialectica* and Talon's *Rhetorica*, in 1584. Richard Schilders, who was also responsible for the second edition in 1588, published Fenner's book. The book must have sold reasonably well because it was reprinted, but I do not know whether the book had any impact on the Dutch

community in Middelburg. On Fenner, see *Nieuw Nederlandsch Biografisch Woordenboek*, vol. 9, col. 254; Howell, *Logic and Rhetoric*, 219-20. On Ramus and Stevin, see J. J. Verdonk, "Vom Einfluss des P. de la Ramée auf Simon Stevin," *Centaurus* 13 (1969): 251-62.

94. In all biographies and studies of Arminius, his Ramism is highlighted. See J. H. Maronier, *Arminius: Een biografie* (Amsterdam: Rogge, 1905); Bangs, *Arminius*; E. Dekker, *Rijker dan Midas: Vrijheid, genade en predestinatie in de theologie van Jacobus Armius (1559-1609)* (Zoetermeer: Uitgeverij Boekencentrum, 1993), esp. 23.

95. Bangs, *Arminius*, 49.

96. Arminius left Leiden in 1581 and went to Geneva to continue his study of theology with Beza. In Geneva, Arminius actively promoted Ramism, as he did in Basel, where he arrived in 1582. Bangs, *Arminius*, 71; Dekker, *Rijker dan Midas*, 23; Hotson, *Commonplace Learning*, 23. Arminius returned to Holland in 1587 to become a minister in Amsterdam. He was appointed as professor of theology at Leiden University in 1603. After leaving Bazel in 1583, Arminius never promoted Ramism, but "the spirit of Ramism lived on in Arminius," so Maronier, *Arminius* tells us.

97. In the early modern period, a "professor extraordinarius" earned less than a full professor (professor ordinarius) and was not a member of the academic senate.

98. For a list of Rudolph Snellius's publications, see Klaas van Berkel, "De geschriften van Rudolf Snellius: Een bijdrage tot de geschiedenis van het ramisme in Nederland," *Tijdschrift voor de Geschiedenis der Geneeskunde, Natuurwetenschappen, Wiskunde en Techniek* 6 (1984): 185-94.

99. "Sine sedula Ramae et Aristotelae philosophiae collatione et coniunctione nemo perfectus hodie philosophus evadere potest." Rudolph Snellius, *Snellio-Ramaeum philosophiae syntagma* (Frankfurt: Johannes Saurius, 1596), 153. See also E. Sellberg, *Filosofin och nyttan I: Petrus Ramus och ramismen* (Göteborg: Acta Universitatis Gothoburgensis, 1979), 131.

100. Snellius, *Syntagma*, 76.

101. Hotson, *Commonplace Learning*, 55-58.

102. H. J. Witkam, *De dagelijkse zaken van de Leidse universiteit van 1581 tot 1596*, 14 vols. (Leiden: H. J. Witkam, 1970-75), 5:88.

103. For the judgment of Lipsius, see P. C. Molhuysen, *Bronnen tot de geschiedenis der Leidsche universiteit*, 7 vols. (The Hague: Martinus Nijhoff, 1913-24), 1:152*-163*; Paul Dibon, *L'enseignement philosophique dans les universités néerlandaises à l'époque pre-cartésienne (1575-1650)* (Leiden: printed by the author, 1954), 73-74.

104. Ibid., 51.

105. In the early seventeenth century, the University of Franeker in Friesland was the real center of Ramism in the Dutch Republic. Klaas van Berkel, "Franeker als centrum van ramisme," in G. Th. Jensma et al., eds., *Universiteit te Franeker 1585-1811: Bijdragen tot de geschiedenis van de Friese hogeschool* (Leeuwarden: Fryske Akademy, 1985), 424-37.

106. Jan vanden Broecke, *Instructie der zee-vaert door de gheheele werelt: Hier by is oock een aenhangsel, om de astrolaby catholicum te leeren trecken . . . Voorts, soo komt hier het voornaemste, dat eenen ingenieur ende landt-meter van noode is* (Rotterdam: Abraham Migoen, 1610).

107. Vanden Broecke's *Instructie der zee-vaert* does not appear in the auction catalog of Beeckman's library; however, he owned several other mathematical textbooks, among others the famous *Cijfferinghe* by Willem Bartjens (1604).

108. *Journal*, 1:6.

109. Ibid., 4:17-19.

110. Other ways to divide the mathematical disciplines were those ascribed to Pythagoras (the four disciplines constituting the quadrivium) and the one proposed by Geminus (the mathematics of the *intelligibilia* against the mathematics of the *sensibilia*).

111. For most of these authors, see *Dictionary of Scientific Biography*. Oronce Finé (1494–1555), a professor of mathematics in Paris from 1532 until his death, was Ramus's teacher and colleague. Henricus Glareanus (1488–1563) ended his career as a professor of history and poetics in Freiburg im Breisgau, but he is better known as the writer of a handbook of music, *Dodecachordon* (1547).

112. *Journal*, 2:405.

113. For the work of Reisner, see Verdonk, *Petrus Ramus*, 71–72, 421; for Pena, see ibid., 64–66, 431–32.

114. *Journal*, 4:26, where van Laren writes, "Ramei nonnuli, quos concedo multos esse judiciosissimos," thus acknowledging that Beeckman was right in calling some Ramist logicians "most judicious."

115. "Agnosces sterilitatem philosophicam" (You are aware of the sterility of philosophy). Ibid., 4:31.

116. Ibid., 1:132–33. See also ibid., 4:41.

117. Ibid., 2:11.

118. "Leges artium, a Rameis tam studiose explicatas" (The rules of the art, so thoroughly explicated by the Ramists) (ibid., 2:313).

119. Ibid., 2:214–15, 217. In Rotterdam, Beeckman also consulted the work of the German Ramist theologian Johannes Piscator (ibid., 2:195; Beeckman writes "a piscatore," which De Waard corrected and replaced by "a praedicatore," apparently unaware that there had been a theologian called Piscator). In the same note, Beeckman also refers to yet another Ramist compendium, Alsted's *Logicae systema harmonicum . . . ex authoribus Peripateticis juxta et Rameis traditur* (Herborn: Christoph Corvinus, 1614).

120. *Journal*, 3:2.

121. Ibid., 4:146.

122. Ibid., 4:157: "Nec de veritate axiomatis Aristotelici admodum sollicitus, qui sciam artes omnes a rusticis doceri et per rusticos probari. Nihil igitur plebs a peritis discit quam id quod periti in plebe se animadvertisse existimant, et dispersa in plebe, male interdum collegerunt. Aures quidem et vocem musicis plebs accommodat, at sibi relicta, tandem frequenti usu edocta, quod bonum est retinet, mala in bonum convertit." This is also why Beeckman was so tolerant toward Cornelis Drebbel. He highly respected Drebbel's mechanical ingenuity and quietly replaced Drebbel's vitalistic (in his eyes, also weird) interpretations by purely mechanical explanations. The same mechanism explains why Ramists especially were so sympathetic to Drebbel's inventions and interpretations. According to Vera Keller, "How to Become a Seventeenth-Century Natural Philosopher: The Case of Cornelis Drebbel (1572–1633)," in Sven Dupré and Christoph Lüthy, eds., *Silent Messengers: The Circulation of Material Objects of Knowledge in the Early Modern Low Countries* (Berlin: LIT Verlag, 2011), 125–51, esp. 137: "A Ramist preference for knowledge found in and for use can help explain the enthusiastic reception of Drebbel's work."

123. Canone, "Il *Catalogus librorum* di Isaac Beeckman," 131–59. Surprisingly, Ramus's books on mathematics are not mentioned in the catalog, and Beeckman's teacher Snellius is represented only by a book on Ramus's *Ethica*.

124. Ong, *Ramus*, 8–9.

125. Ibid., 115. Ong was inclined to look for the origin of the mechanistic tendency in Ramism in late medieval scholastic thinking. As one of his reviewers put it, "The origin of the movement [that is, Ramism] is located in the practical requirements of the pedagogical situation in European universities, particularly in the necessity of making logic intelligible for the children who composed the student bodies. Ramism is shown to have developed a tendency in scholastic logic to identify knowledge with teaching, and teaching with a simplified spatial approach to reality, a tendency which was reinforced by the diagrammatic tidiness made possible by letterpress printing. The Ramist outlook is an extension of the mechanical bias of the scholastic mind" (T. K. Scott Jr., review of Ong, *Ramus*, *Journal of Philosophy* 59 [1962]: 556–57).

126. *Journal*, 2:102.

127. Van Laren to Beeckman, October 12, 1613 (ibid., 4:26–28). The question at stake was whether "in logic an indivuum has to be labeled as a species" (an individuum in Logica debeat appellaris species) (ibid., 4:26). Beeckman defended this thesis, partly on the basis of the "common notion" (notio communis) (or actually Ockham's razor) "Quod fieri potest per pauciora, male fit per plura." However, it is clear from what follows that he had in fact cited several Ramists ("Ramei nonnulli"), who were very judicious in these matters, as even the Aristotelian Van Laren acknowledged. See also note 115.

128. Ibid., 1:10.

129. Ibid., 4:147: Commenting on his principle of inertia, Beeckman writes to Mersenne in June 1629: "Nothing more certain has ever come to my mind and during [the past] *twenty years* I have not read, heard or contemplated anything that would have made me doubt this even a little bit" (Qua ratione nihil unquam certius in mentem mihi venit, nec viginti annis quicquam legi, audivi aut meditatus sum quod minimam erroris suspicionem mihi hic movere potuerit") (emphasis added).

130. Ibid., 1:345.

131. Ibid., 4:122–26.

132. Verdonk, *Petrus Ramus*, 96–97.

133. Ibid., 293.

134. A possible reason for not mentioning Ramus may have been the introduction of a new school program in the province of Holland just a few years before Beeckman was appointed in Dordrecht. This new program—drafted by the humanist Gerard Vossius, a former rector at Dordrecht, and a professor at Leiden in 1625—was intended to supplant Ramist influences in favor of more traditional humanist learning.

135. *Journal*, 3:298.

136. Ibid., 1:26; 2:102.

137. Alexandre Koyré, *Études galiléennes* (Paris: Hermann, 1939; reprint, 1966), 15.

CHAPTER 7: BEECKMAN AND THE SCIENTIFIC REVOLUTION

1. E. J. Dijksterhuis, *The Mechanization of the World Picture* (Oxford: Oxford University Press, 1961), 330.

2. Ibid.

3. "Il semble que le premier initiateur, jusqu'ici méconnu, d'un grand nombre des idées du XVIIe siècle ait été le modeste et silencieux Beeckman." Robert Lenoble, *Mersenne ou la naissance du mécanisme* (Paris: Vrin, 1943; reprint, 2003), 427.

4. *Journal*, 1:59.

5. Ibid., 3:123; 4:153, 189.

6. Benedino Gemelli, *Isaac Beeckman: Atomista e lettore critico di Lucrezio* (Florence: Leo S. Olschki, 2002), ix, 12.

7. See chapter 4.

8. Cf. B. Rochot, "Beeckman, Gassendi et le principe de l'inertie," *Archives Internationales d'Histoire des Sciences* 31 (1952): 282–89; P. A. Pav, "Gassendi's Statement of the Principle of Inertia," *Isis* 57 (1966): 24–34.

9. "Quocirca non uni, sed tribus minimum amicis haec tradenda, nec nimis temere desperandum." *Journal*, 2:377.

10. In a footnote, De Waard states that Beeckman had "without doubt" (sans doute) shown his notes to Descartes in 1618. However, it is more likely that Beeckman allowed Descartes to consult his notebooks only upon the renewal of their friendship in 1628 and 1629. On Hortensius, see ibid., 3:354.

11. Rienk Vermij, *The Calvinist Copernicans: The Reception of the New Astronomy in the Dutch Republic, 1575–1750* (Amsterdam: Edita, 2002), 73–92.

12. Klaas van Berkel, "De illusies van Martinus Hortensius: Natuurwetenschap en patronage in de Republiek," in Van Berkel, *Citaten uit het boek der natuur. Opstellen over wetenschapsgeschiedenis* (Amsterdam: Bert Bakker, 1998), 63–84; Vermij, *The Calvinist Copernicans*, 126–29.

13. *Journal*, 4:207, n. 2; *Correspondance du P. Marin Mersenne religieux minime*, 17 vols. (Paris: Beauchesne, 1932–88), 3:205; 3:407–8. Cf. *Journal*, 4:214, n. 5.

14. "La rencontre de Descartes et de Beeckman en 1618 a une extrême importance dans l'histoire de la pensée du père du cartesianisme. Mais elle est simple cause occasionelle, et ne laisse nulle trace, ne porte aucun reflet sur la philosophie cartésienne." Étienne Souriau, *L'instauration philosophique* (Paris: Librairie Félix Alcan, 1939), 58, n. 1.

15. "En effet, Beeckman, on s'en rend compte maintenant, mérite pleinement l'appellation de *vir ingeniosissimus* dont l'avait gratifié Descartes; et, ce qui plus est, il nous apparaît desormais comme un chaînon de première importance dans l'histoire de l'évolution des idées scientifiques; enfin, son influence sur Descartes semble avoir été beaucoup plus profonde que l'on n'a pu le supposer jusqu'ici." Alexandre Koyré, *Études galiléennes* (Paris: Hermann, 1939; reprint, 1966), 108–9, n. 2.

16. John Schuster, *Descartes and the Scientific Revolution, 1618–1634: An Interpretation* (Ann Arbor: University Microfilms International, 1977), 51–52.

17. Ibid., 69–70.

18. Stephen Gaukroger and John Schuster, "The Hydrostatic Paradox and the Origins of Cartesian Dynamics," *Studies in the History and Philosophy of Science* 33 (2002): 535–72, esp. 558.

19. Cf. Klaas van Berkel, "Beeckman, Descartes et la philosophie physico-mathématique," *Archives de Philosophie* 46 (1983): 620–26.

20. Richard Arthur, "Beeckman, Descartes and the Force of Motion," *Journal of the History of Philosophy* 45 (2007): 1–28, esp. 3.

21. Ibid., 15, n. 36.

22. Ibid., 27.

23. Schuster, *Descartes*, 373–87; Gaukroger and Schuster, "Hydrostatic Paradox."

24. There were several reasons why Beeckman—who in 1626 had decided *not* to publish his ideas—changed his mind. Most significant was the sense of confidence he had acquired

in the more distinguished cultural setting in Dordrecht. Another reason may have been that the Dutch Republic had regained the initiative on the battlefield. The new stadholder of Holland, Frederick Henry, was on the offensive and, in September 1629, took the city of 's Hertogenbosch, some thirty kilometers from Dordrecht (during his visit to Holland, Gassendi went to see the siege). Finally, Beeckman's review of Kepler's ideas prompted Beeckman to feel that his ideas were at least as good as those of the well-known German astronomer and therefore worthwhile publishing after all.

25. Schuster, *Descartes*, 565–79, 590–93.

26. Fearing that Ferrier might be able to construct lenses without his help, Descartes began to picture Ferrier as a simple craftsman, whose work would be worthless without the guidance of the mathematician Descartes. "Even if I described in detail the machines required to make the lens," Descartes wrote to Mersenne on April 15, 1630, "I do not think that he can do without me." Quoted in William R. Shea, "Descartes and the French Artisan Jean Ferrier," *Annali dell'Istituto e Museo di Storia della Scienza di Firenze* 7 (1982): 145–60, esp. 155.

27. Klaas van Berkel, "Descartes' Debt to Beeckman: Inspiration, Cooperation, Conflict," in Stephen Gaukroger et al., eds., *Descartes' Natural Philosophy* (London: Routledge, 2000), 46–59. However, the German philosopher G.W. Leibniz became suspicious when he later read Adrien Baillet's biography of Descartes in which Beeckman was belittled. "It seems to me that people have been unjust to Monsieur Isaac Beeckman, by mistreating him on the basis of the reports in the letters of Monsieur Descartes only, from which I have learned not to boast at the expense of others, because Monsieur Descartes gives a strange twist to things when he was mad at someone" (Il me semble qu'on fait tort à M. Isaac Beckman en le maltraittant sur le seul rapport des lettres de M. des Cartes, aux quelles j'ay appris qu'on ne se doit point fier au desavantage des gens, car M. des Cartes donnoit un étrange tour aux choses quand il estoit piqué contre quelcun). As quoted in Gemelli, *Beeckman*, 127, n. 11.

28. J. A. Worp, ed., *De briefwisseling van Constantijn Huygens (1608–1687)*, 6 vols. (The Hague: Martinus Nijhoff, 1911–17), 4:47 (the book that Worp, the editor of Huygens's correspondence, is referring to elsewhere [ibid., 4:42] and identifies as Beeckman's *Centuria* is in fact Mersenne's *Cogitata physico mathematica*, also published in 1644). On Smith, see ibid., 1:433. Because one of the last notes in the *Centuria* was dated 1629, Smith must have concluded that the book had been completed by 1628. Three of the four entries mentioned by Smith indeed deal with magnetism (#36 = *Journal*, 1:36; #88 = ibid., 3:17; #83(b) = ibid., 3:26), while the fourth (#77 = ibid., 3:123) is a note on Gassendi's visit (magnetic corpuscles are mentioned in passing).

29. Dijksterhuis, *Mechanization*, 495–501. For an analysis of the background of Dijksterhuis's conception of mechanization, see Klaas van Berkel, *E. J. Dijksterhuis: Een biografie* (Amsterdam: Bert Bakker, 1996).

30. Dijksterhuis, *Mechanization*, 496.

31. Ibid.

32. Ibid., 497.

33. I owe the reference to Fludd's emblematic pictures to Christoph Lüthy.

34. *Œuvres complètes de Christiaan Huygens*, 22 vols. (The Hague: Martinus Nijhoff, 1888–1950), 21:471; G.W. Leibniz in *Acta Eruditorum* (1710), 412. See A. Rupert Hall, *Philosophers at War: The Quarrel between Newton and Leibniz* (Cambridge: Cambridge University Press, 1980), 146–67, esp. 162–63.

35. Dijksterhuis, *Mechanization*, 501.

36. Renzo Baldazzo, "The Role of Visual Representation in the Scientific Revolution: A Historiographical Inquiry," *Centaurus* 48 (2006): 69–88.

37. Therefore, William Ashworth wrote in 1991 that "we [the historians of science] paid virtually no attention to the visual side of the scientific revolution." William Ashworth, "The Scientific Revolution: The Problem of Visual Authority," in *Conference on Critical Problems and Research Frontiers in History of Science and History of Technology* (Madison: University of Wisconsin Press, 1991), 326–48), 326, quoted in Baldazzo, "Role of Visual Representation," 87–88.

38. Cf. Harold J. Cook, *Matters of Exchange: Commerce, Medicine, and Science in the Dutch Golden Age* (New Haven: Yale University Press, 2007). Cook directly relates the new stress on accurate description and the correct identification of natural particulars, so important in early modern natural history and medicine, to the values of the merchant elite of Europe in this period, particularly those merchants engaged in overseas trade with the East and West Indies.

39. Christoph Lüthy, "Where Logical Necessity Becomes Visual Persuasion: Descartes' Clear and Distinct Illustrations," in Sachiko Kusukawa and Ian Maclean, eds., *Transmitting Knowledge: Words, Images, and Instruments in Early Modern Europe* (Oxford: Oxford University Press, 2006), 97–133. See also Brian S. Baigrie, "Descartes's Scientific Illustrations and 'la grande mécanique de la nature,'" in Brian S. Baigrie, *Picturing Knowledge: Historical and Philosophical Problems concerning the Use of Art in Science* (Toronto: Toronto University Press, 1996), 86–134.

40. Quoted in John Henry, "Occult Qualities and the Experimental Philosophy: Active Principles in Pre-Newtonian Matter Theory," *History of Science* 24 (1986): 335–81, esp. 335. Petty was also a physician, political economist, and one of the founders of the Royal Society.

41. For the drawing of the microscope, see *Journal*, 3:442 (and see fig. 2).

42. Svetlana Alpers, *The Art of Describing: Dutch Art in the Seventeenth Century* (Chicago: University of Chicago Press, 1983).

43. Ibid., 76.

44. David A. Freedberg and Jan de Vries, eds., *Art in History, History in Art: Studies in Seventeenth Century Dutch Culture* (Santa Monica, CA: Getty Center for the History of Art and the Humanities, 1991); Wayne Franits, ed., *Looking at Seventeenth-Century Dutch Art: Realism Reconsidered* (Cambridge: Cambridge University Press, 1997). With regard to Ruisdael's "billowing clouds," see J. Walsh, "Skies and Reality in Dutch Landscape," in Freedberg and De Vries, *Art in History*, 95–117. Walsh's conclusion is clear: although the rendering of clouds by Ruisdael and others had its basis in nature, the paintings were far from "realistic," photographic images of the skies over Holland. "Selecting, stereotyping, and altering for the sake of more effective images of nature, all resemble the processes of flower painters, who combined blossoms from different times of the year in impossible bouquets, or view painters, who put well-known monuments from different cities into the same picture" (ibid., 109–10). Genre paintings (representing scenes from everyday life) were likewise definitely not meant to be expressions of real everyday life. They were valued for the way they told a moral or simply funny story to the onlooker as well as for a quality that contemporaries circumscribed as *schilderachtig* (picturesque). See B. Bakker, "*Schilderachtig.* Discussions of a Seventeenth-Century Term and Concept," *Simiolus* 23 (1995): 147–62; Lyckle de Vries, *Verhalen uit kamer,*

keuken en kroeg: Het Hollandse genre van de zeventiende eeuw als vertellende schilderkunst (Amsterdam: Amsterdam University Press, 2005). I thank Lyckle de Vries for these references and for discussing them with me.

45. Mariët Westermann, *A Worldly Art: The Dutch Republic, 1585–1718* (New York: Abrams, 1996), 90–91.

46. On Bosschaert: L. J. Bol, *The Bosschaert Dynasty: Painters of Flowers and Fruit* (Leigh-on-Sea: Lewis, 1960), 14–33; Noortje Bakker et al., eds., *Masters of Middelburg: Exhibition in the Honour of Laurens J. Bol* (Amsterdam: Kunsthandel K. & V. Waterman, 1984); Klaas van Berkel, "The City of Middelburg, Cradle of the Telescope," in Albert van Helden et al., eds., *The Origins of the Telescope* (Amsterdam: KNAW Press, 2010), 45–71. Shortly before 1615, Bosschaert moved to Bergen op Zoom and later to Utrecht and Breda.

47. Westermann, *A Worldly Art*, 88.

48. Pamela H. Smith, *The Body of the Artisan: Art and Experience in the Scientific Revolution* (Chicago: University of Chicago Press, 2004). See also Eric Jorink and Bart Ramakers, eds., *Art and Science in the Early Modern Netherlands* (Zwolle: Wbooks, 2011).

49. Anne Goldgar, *Tulipmania: Money, Honor, and Knowledge in the Dutch Golden Age* (Chicago: University of Chicago Press, 2007).

50. Van Berkel, "City of Middelburg," 61–71; Florike Egmond, *The World of Carolus Clusius: Natural History in the Making, 1550–1610* (London: Routledge, 2010), 143–56.

51. *Journal*, 1:xxii; 4:49, 283. De Waard in his footnote to p. 49 doubts whether the "Caspar" mentioned by Beeckman as his co-traveler in 1618 really is the same man as the Caspar Adriaansz. Parduyn, who would later succeed him as principal of the Latin school in Dordrecht. Beeckman must have referred to Caspar Parduyn, the son of the Middelburg merchant and amateur botanist Willem Jaspersz. Parduyn (a correspondent of Clusius).

Bibliographic Essay

The obvious place to begin any research on Isaac Beeckman's natural philosophy is the *Journal tenu par Isaac Beeckman de 1604 à 1634, publié avec une introduction et des notes par C. de Waard*, 4 vols. (The Hague: Martinus Nijhoff, 1939–53). This excellent edition includes scientific and technical notes as well as remarks of a more personal nature. In addition to Beeckman's correspondence, the fourth volume comprises almost all archival material concerning his life and that of his relatives, most of which was destroyed during the Second World War. The manuscript of Beeckman's notebooks was also damaged during the war, but has been restored almost completely (Middelburg, Zeeuwse Bibliotheek, ms. no. 6471). The *Journal* is available on the Internet in two editions: www .historyofscience.nl and www.dbnl.nl/auteurs/beeckman. Beeckman's correspondence is also partially published in the major editions of the correspondences of René Descartes and Marin Mersenne: *Oeuvres de Descartes*, publiées par Charles Adam et Paul Tannery, 13 vols. (Paris: Vrin, 1964–75), vol. 1 (originally published 1897); *Correspondance du P. Marin Mersenne religieux minime*, publiée par Mme Paul Tannery, éditée et annotée par Cornelis de Waard et al., 17 vols. (Paris: Beauchesne, 1932–88), vols. 2–6. During his lifetime Beeckman published only his dissertation, *Theses de Febre Tertiana* (Cadomi [Caen]: Ex typographia Jacobi Bassi, 1618), of which only the British Museum owns a partial copy (reproduced in the *Journal*, 4:42–44). Abraham published a selection of his brother Isaac's notes some years after Beeckman's death as his *Mathematicophysicarum Meditationum, Quaestionum, Solutionum Centuria* (Utrecht: Petrus Daniel Sloot, 1644). An interesting source on Beeckman's intellectual outlook is the auction catalog of his library, reproduced by Eugenio Canone, "Il *Catalogus librorum* di Isaac Beeckman," *Nouvelles de la Republique des Lettres* 1991, no. 1 (1992): 131–59.

Cornelis de Waard offers a very factual biography of Beeckman in the first volume of his edition of the notebooks: "Vie de l'auteur," *Journal* (1939), 1:i–xxiv. This was the basis for the intellectual portrait De Waard sketched in "Isaac Beeckman (1588–1637)," *Archives du Musée Teyler*, ser. 3, 9 (1947): 299–342. Klaas van Berkel's *Isaac Beeckman (1588–1637) en de mechanisering van het wereldbeeld* (Amsterdam: Rodopi, 1983) is the first book-length treatment of Beeckman's life and work. It forms the basis for the present volume. Klaas van Berkel, "Isaac Beeckman," in *Ueberweg Grundriss der Geschichte der Philosophie: Die Philosophie des 17. Jahrhunderts; Frankreich und Niederlande*, Zweiter Halbband (Basel: Schwabe, 1993), 631–36, includes new material.

The secondary literature concerning Beeckman's ideas on natural philosophy is highly fragmented and consists mainly of articles devoted to special aspects of his scientific work. De Waard announced the discovery of Beeckman's notebooks in "Eene correspondentie van Descartes uit de jaren 1618 en 1619," *Nieuw Archief voor Wiskunde*, Tweede Reeks, 7 (1905): 69–87, focusing our attention from the very start on Beeckman's relationship with Descartes, as if he was only worthwhile studying in the context of the genesis of Descartes's natural philosophy. De Waard further analyzed Beeckman's thoughts on the rules of collision in his "Sur les règles du choc des corps d'après Beeckman," *Correspondence de Mersenne*, 2:632–44. E. J. Dijksterhuis discusses Beeckman's meaning for the science of mechanics in *The Mechanization of the World Picture*, trans. C. Dikshoorn (Oxford: Oxford University Press, 1961), 329–33. More recent discussions of Beeckman's theory of mechanics and its presumed influence on Descartes include Peter Damerow and Gideon Freudenthal, *Exploring the Limits of Preclassical Mechanics: A Study of Conceptual Development in Early Modern Science; Free Fall and Compounded Motion in the Work of Descartes, Galileo, and Beeckman* (New York: Springer, 2004 [1991]); and Richard Arthur, "Beeckman, Descartes and the Force of Motion," *Journal of the History of Philosophy* 45 (2007): 1–28.

Besides mechanics, several other topics discussed by Beeckman have received the attention of scholars and historians. De Waard discusses the opinions of Beeckman and several of his contemporaries regarding air pressure in his *L'expérience barométrique. Ses antécédents et ses explications* (Thouars (Deux Sèvres): Imprimerie nouvelle, 1936). Giancarlo Nonnoi, *Il pelago d'aria. Galileo, Baliani, Beeckman* (Rome: Bulzoni Editore, 1988), covers the same ground. The way Reyer Hooykaas deals with the religious dimension of Beeckman's work in his "Science and Religion in the Seventeenth Century: Isaac Beeckman, 1588–1637," *Free University Quarterly* 1 (1951): 169–83, is instructive but also somewhat schematic. H. H. Kubbinga focuses entirely on Beeckman's ideas on the structure of matter in his rather finalistic *L'histoire du concept de "molecule,"* 3 vols. (Paris: Springer, 2002), 1:203–37. Benedino Gemelli offers a very detailed discussion of the influence of the Roman poet Lucretius on Beeckman in his *Isaac Beeckman: Atomista e lettore critico di Lucrezio* (Florence: Leo S. Olschki, 2002). Few historians have as yet paid attention to Beeckman's extensive writings on medicine, except for M. J. van Lieburg, "Isaac Beeckman (1588–1637) and His Diary-Notes on William Harvey's Theory on Blood Circulation (1633–1634)," *Janus* 69 (1982): 161–83. Elisabeth Moreau, "Le substrat galénique des idées médicales d'Isaac Beeckman (1616–1627)," *Studium* 4 (2011): 137–51, begins to fill the gap. Studies on Beeckman's minor role in the history of the telescope include Cornelis de Waard, *De uitvinding der verrekijkers: eene bijdrage tot de beschavingsgeschiedenis* (The Hague: De Nederlandsche Boek- en Steendrukkerij, 1906); Albert van Helden, *The Invention of the Telescope* (Philadelphia: American Philosophical Society, 1977); and Fokko Jan Dijksterhuis, "Labour on Lenses: Isaac Beeckman's Notes on Lens Making," in Albert van Helden et al., eds., *The Origins of the Telescope* (Amsterdam: KNAW Press, 2010), 257–70. Finally, H. Floris Cohen provides a lucid analysis of Beeckman's ideas on music in his *Quantifying Music: The Science of Music at the First Stage of the Scientific Revolution, 1580–1650* (Dordrecht: Reidel, 1984). This analysis should be supplemented by Frédéric de Buzon, "Science de la nature et théorie musicale chez

Isaac Beeckman (1588–1637)," *Revue d'Histoire des Sciences* 38 (1985): 97–120. De Buzon also produced a fine edition of Descartes's *Abrégé de musique. Compendium musicae* (Paris: Presses universitaires de France, 1987). For a summary of much of the secondary literature, see Eio Honma, "Beeckman's Natural Philosophy," *Scientiarum Historia* (Japan) 5 (1996): 225–47.

The best introduction to the wider historical context of Beeckman's life in the Dutch Republic is Jonathan I. Israel, *The Dutch Republic: Its Rise, Greatness, and Fall, 1477–1806* (Oxford: Clarendon Press, 1995). Much shorter but no less authoritative is Maarten Prak, *The Dutch Republic in the Seventeenth Century: The Golden Age* (Cambridge: Cambridge University Press, 2005). Finally several chapters in Willem Frijhoff and Marijke Spies, *1650: Hard-Won Unity* (Assen / Basingstoke: Royal Van Gorcum / Palgrave Macmillan, 2004), can also be consulted with profit.

Beeckman's parents moved from the southern Netherlands to the Dutch Republic, together with thousands of others. According to the standard view, this massive immigration around 1600 resulted in a sudden rise of science and scholarship in the Dutch Republic. See J. G. C. A. Briels, *Zuidnederlandse immigratie 1572–1630* (Haarlem: Fibula–Van Dishoeck, 1978). Angelo De Bruycker and Djoeke van Netten, " 'Zodat mijn verbanning tegelijk jouw straf is.' Bloei, verval en migratie van wetenschap in de Republiek en de Spaanse Nederlanden," *Bijdragen en Mededelingen betreffende de Geschiedenis der Nederlanden / Low Countries Historical Review* 123 (2008): 3–30, offer a critical reevaluation of the standard view. Helpful in this respect is also Oscar Gelderblom, *Zuid-Nederlandse kooplieden en de opkomst van de Amsterdamse stapelmarkt (1578–1630)* (Hilversum: Verloren, 2000).

For a general introduction to the history of science in the Netherlands in the early modern period, one should start with K. van Berkel, L. C. Palm, and Albert van Helden, eds., *The History of Science in the Netherlands: Survey, Themes, and Reference* (Leiden: Brill, 1999), esp. 13–94. Older, but still very useful is Dirk Struik's more essayistic *The Land of Stevin and Huygens: A Sketch of Science and Technology in the Dutch Republic during the Golden Century* (Dordrecht: Reidel, 1981), a translation of a Dutch monograph from 1947. For a reevaluation of the nature of scientific life in seventeenth-century Holland, see Eric Jorink, *Wereldbeeld en wetenschap in de Gouden Eeuw* (Hilversum: Verloren, 1999), and, by the same author, *Reading the Book of Nature in the Dutch Golden Age, 1575–1715* (Leiden: Brill, 2010). The author stresses the importance of natural history over more traditional topics like mechanics and engineering. On a much broader canvas, Harold J. Cook, *Matters of Exchange: Commerce, Medicine, and Science in the Dutch Golden Age* (New Haven: Yale University Press, 2007), does the same. For a critical review, see Klaas van Berkel, "Rediscovering Clusius: How Dutch Commerce Contributed to the Emergence of Modern Science," *Bijdragen en Mededelingen betreffende de Geschiedenis der Nederlanden / Low Countries Historical Review* 123 (2008): 227–36.

Studies on special themes in the history of science in the Dutch Republic with some relevance to the context in which Beeckman operated include C. S. Maffioli and L. C. Palm, eds., *Italian Scientists in the Low Countries in the XVIIth and XVIIIth Centuries*

(Amsterdam: Rodopi, 1989); Tabitta van Nouhuys, *The Age of Two-Faced Janus: The Comets of 1577 and 1618 and the Decline of the Aristotelian World View in the Netherlands* (Leiden: Brill, 1998); Rienk Vermij, *The Calvinist Copernicans: The Reception of the New Astronomy in the Dutch Republic, 1575–1750* (Amsterdam: Edita, 2002), with a section on Beeckman on pp. 113–18; Karel Davids, *The Rise and Decline of Dutch Technological Leadership: Technology, Economy and Culture in the Netherlands, 1350–1800*, 2 vols. (Leiden: Brill, 2008); Florike Egmond, *The World of Carolus Clusius: Natural History in the Making, 1550–1610* (London: Pickering & Chatto, 2010); and Christoph Lüthy, *David Gorlaeus (1591–1612): An Enigmatic Figure in the History of Philosophy and Science* (Amsterdam: Amsterdam University Press, 2012).

_ৎ ৎ_

Beeckman's natural philosophy is part of the scientific revolution of the seventeenth century. Although most historians of science have discarded the notion itself, few of them deny that there was a fundamental transformation of the study of nature in the early modern period. For practical purposes, historians continue to talk about the "scientific revolution," and the literature on it, already extensive, is still expanding rapidly. H. Floris Cohen, *The Scientific Revolution: A Historiographical Inquiry* (Chicago: University of Chicago Press, 1994) offers a comprehensive survey of the historical tradition. Major surveys of the topic include A. Rupert Hall, *The Scientific Revolution, 1500–1800: The Formation of the Modern Scientific Attitude*, 2nd and rev. ed. (London: Longman, 1962), re-edited as *The Revolution in Science, 1500–1750* (London: Longman, 1983); Richard S. Westfall, *The Construction of Modern Science: Mechanisms and Mechanics* (Cambridge: Cambridge University Press, 1977 [1971]); John Henry, *The Scientific Revolution and the Origins of Modern Science* (Basingstoke: Macmillan, 1997); and Stephen Gaukroger, *The Emergence of a Scientific Culture: Science and the Shaping of Modernity, 1210–1685* (Oxford: Clarendon Press, 2006). Peter Dear, *Revolutionizing the Sciences: European Knowledge and Its Ambitions, 1500–1700* (Basingstoke: Palgrave, 2001) is more selective in its approach. Also very useful is the third volume of the authoritative Cambridge History of Science series (ed. David C. Lindberg and Ronald Numbers): Katharine Park and Lorraine Daston, eds., *Early Modern Science* (Cambridge: Cambridge University Press, 2006). Steven Shapin in *The Scientific Revolution* (Chicago: University of Chicago Press, 1996), is somewhat controversial for denying that "the scientific revolution" ever existed, while still writing another book about it.

The founding father of the intellectual interpretation of the scientific revolution is Alexandre Koyré, whose classical *Études galiléennes*, 3 vols. (Paris: Herman, 1939; reprint, 1966) was translated as *Galileo Studies* (Hassocks: Harvester Press, 1978). The book deals extensively with the 1618 exchange between Beeckman and Descartes. Herbert Butterfield's *The Origins of Modern Science: 1300–1800* (London: Bell and Hyman, 1949) also sees the rise of the new science mainly as a transformation of intellectual thought. E. J. Dijksterhuis, *The Mechanization of the World Picture* is another widely read but mainly intellectual interpretation of what went on in science in the early modern period (the author denies that it was a revolution). H. Floris Cohen, *How Modern Science Came into the World: Four Civilizations, One 17th-Century Breakthrough* (Amsterdam:

Amsterdam University Press, 2010), is a modern defense of the intellectual approach to early modern science. The major alternative to the Koyré tradition is the social interpretation of the origins of modern science. Edgar Zilsel was the first serious historian of science to study early modern science in this way. His essays are collected as *Die sozialen Ursprünge der neuzeitlichen Wissenschaft,* edited and partially translated by Wolfgang Krohn (Frankfurt am Main: Suhrkamp, 1976). For the English edition, see Diederik Raven, Wolfgang Krohn, and Robert S. Cohen, eds., *The Social Origins of Modern Science* (Dordrecht: Kluwer Academic Publishers, 2000). Paolo Rossi developed a comparable approach to early modern science in *Philosophy, Technology, and the Arts in the Early Modern Era,* trans. Salvator Attanasio (New York: Harper & Row 1970), originally published in Italian in 1962. A modern exponent of the view that the input of artisanal knowledge was instrumental in bringing forward modern science is Pamela S. Smith, *The Body of the Artisan: Art and Experience in the Scientific Revolution* (Chicago: University of Chicago Press, 2004). Lissa Roberts, Simon Schaffer, and Peter Dear, eds., *The Mindful Hand: Inquiry and Invention from the Late Renaissance to Early Industrialization* (Amsterdam: Edita, 2007), is excellent for its theoretical approach. Klaas van Berkel, "The Dutch Republic: Laboratory of the Scientific Revolution," in Klaas van Berkel and Leonie de Goei, eds., *The International Relevance of Dutch History,* special issue of the *Bijdragen en Mededelingen betreffende de Geschiedenis der Nederlanden / Low Countries Historical Review* 125 (2010): 81–105, argues that, because of the overwhelmingly practical and nonmetaphysical nature of Dutch science in the early modern era, the Netherlands is the ideal site for studying the mechanisms behind the rise of the new science.

There was a time that the scientific revolution was more or less equivalent to what was called the mechanization of the worldview. This concept was introduced by Anneliese Maier in her short study *Die Mechanisierung des Weltbilds im 17. Jahrhundert* (Leipzig: Felix Meiner, 1938), reprinted in Anneliese Maier, *Zwei Untersuchungen zur nachscholastischen Philosophie: Die Mechanisierung des Weltbildes im 17. Jahrhundert, Kants Qualitätskategorien* (Rome: Edizion di Storia e Letteratura, 1968). Dijksterhuis's *The Mechanization of the World Picture* further developed the concept, interpreting the mechanization of the world picture as being essentially a process of mathematization of science. Marie Boas, "The Establishment of the Mechanical Philosophy," *Osiris* 10 (1952): 412–541, offers a somewhat different interpretation, with more stress on corpuscularism and empiricism and less on mathematics and mechanics. Several articles by Alan Gabbey provide an excellent study of the complexities of the notion of a mechanical philosophy. They include "The Mechanical Philosophy and Its Problems," in J. C. Pitt, ed., *Change and Progress in Modern Science* (Dordrecht: Reidel, 1985), 9–84; and "What Was 'Mechanical' about the Mechanical Philosophy?" in Carla Rita Palmerino and J. M. M. Thijssen, eds., *The Reception of the Galilean Science of Motion in Seventeenth-Century Europe* (Dordrecht: Reidel, 2004), 11–23. Jim Bennett, "The Mechanic's Philosophy and the Mechanical Philosophy," *History of Science* 24 (1986): 1–28, is important because it links mechanical philosophy and artisanal practice. A major problem facing the historians of the mechanical philosophy is the growing awareness that there is no clear-cut division between mechanical and vitalistic conceptions of matter. Antonio

Clericuzio's *Elements, Principles and Corpuscles: A Study of Atomism and Chemistry in the Seventeenth Century* (Dordrecht: Reidel, 2000) is indispensable in this context. Christoph Lüthy, John E. Murdoch, and William R. Newman, eds., *Medieval and Early Modern Matter Theories* (Leiden: Brill, 2001), remains the best introduction the topic. Margaret Osler's contribution to this volume, "How Mechanical Was the Mechanical Philosophy?" (423–39), deserves special mention.

Most of Beeckman's predecessors and intellectual peers have inspired more than one biographer. For Simon Stevin, we can still rely on E. J. Dijksterhuis, *Simon Stevin* (The Hague: Martinus Nijhoff, 1943), partly translated as *Simon Stevin: Science in the Netherlands around 1600* (The Hague: Martinus Nijhoff, 1970). Jozef T. Devreeze and Guido Vanden Berghe, "*Wonder en is gheen wonder*": *De geniale wereld van Simon Stevin 1548–1620* (Louvain: Davidsfonds, 2003), offers new material but is also slightly hagiographical. For Beeckman's exact contemporary Marin Mersenne, there is no modern equivalent to the substantial biography by Robert Lenoble, *Mersenne ou la naissance du mécanisme* (Paris: Vrin, 1943; reprint, 2003). Peter Dear, *Mersenne and the Learning of the Schools* (Ithaca: Cornell University Press, 1988), adds however a modern perspective. Among the many studies of Gassendi, mention should be made of Howard Jones, *Pierre Gassendi, 1592–1655: An Intellectual Biography* (Nieuwkoop: B. De Graaf, 1981), and Antonia Lolordo, *Gassendi and the Birth of Early Modern Philosophy* (Cambridge: Cambridge University Press, 2007). Lolordo's book offers the first comprehensive treatment (in English) of Gassendi's philosophical system.

Of all Beeckman's friends and correspondents, Descartes is especially well served by modern scholarship. Authoritative, with strong sections on the Descartes-Beeckman exchange, are Stephen Gaukroger, *Descartes: An Intellectual Biography* (New York: Oxford University Press, 1995), and Desmond M. Clarke, *Descartes: A Biography* (Cambridge: Cambridge University Press, 2006). Geneviève Rodis-Lewis, *Descartes: His Life and Thought* (Ithaca: Cornell University Press, 1998), originally published in French in 1995, offers a more philosophical interpretation of Descartes. William R. Shea, *The Magic of Numbers and Motion: The Scientific Career of René Descartes* (Canton, MA: Watson, 1990), is somewhat narrower in scope than the biographies just mentioned. A. C. Grayling's *Descartes: The Life and Times of a Genius* (New York: Walker, 2005) is highly speculative and ought to be handled with care. The author claims that Descartes acted as a spy for the French during most of his travels abroad. For a discussion of the nature of the relationship between Beeckman and Descartes, see Klaas van Berkel, "Descartes' Debt to Beeckman: Inspiration, Cooperation, Conflict," in Stephen Gaukroger, John Schuster, and John Sutton, eds., *Descartes' Natural Philosophy* (London: Routledge, 2000), 46–59. John Schuster's massive *Descartes-Agonistes: Physico-mathematics, Method and Corpuscular-Mechanism, 1618–1633*, Studies in the History and Philosophy of Science, vol. 27 (Dordrecht: Springer, 2013), based on earlier work on the young Descartes referred to in the notes, was published too late for inclusion in the text of this book. It is, however, the most extensive conceptual and technical discussion of Descartes's first system of natural philosophy. Beeckman is presented as a major source of inspiration for Descartes's corpuscular-mechanical philosophy of nature.

For the imprint Ramism had on European culture and society in the early modern period, see Walter Ong, *Ramus, Method, and the Decay of Dialogue: From the Art of Discourse to the Art of Reason*, with a foreword by Adrian Johns (Chicago: University of Chicago Press, 2004 [1958]). Howard Hotson, *Commonplace Learning: Ramism and Its German Ramifications, 1543–1630* (Oxford: Oxford University Press 2007), is very strong on the social context of Ramism. The two major Ramists in the Netherlands, Rudolph and Willebrord Snellius, the former Beeckman's Leiden teacher, the latter his lifelong friend, form the subject of Liesbeth de Wreede's thesis *Willebrord Snellius (1580–1626): A Humanist Reshaping the Mathematical Sciences* (Utrecht: printed by the author, 2007). The role of illustrations in scientific books and the visual turn in intellectual life in general, discussed and criticized by Ong, has attracted more attention recently. Renzo Baldazzo, "The Role of Visual Representation in the Scientific Revolution: A Historiographical Inquiry," *Centaurus* 48 (2006): 69–88, offers an excellent introduction to the topic and provides a useful bibliography. Also instructive is Christoph Lüthy, "Where Logical Necessity Becomes Visual Persuasion: Descartes' Clear and Distinct Illustrations," in Sachiko Kusukawa and Ian Maclean, eds., *Transmitting Knowledge: Words, Images, and Instruments in Early Modern Europe* (Oxford: Oxford University Press, 2006), 97–133. Svetlana Alpers, *The Art of Describing: Dutch Art in the Seventeenth Century* (Chicago: University of Chicago Press, 1983), was the first to discuss the parallels—real or imagined—between empiricism, visual imagery in science, and realism in contemporary Dutch art. Mariët Westermann, *A Worldly Art: The Dutch Republic, 1585–1718* (New York: Abrams, 1996), presents a more sophisticated argument based on the notion of "reality effect."

Index